SpringerBriefs in Economics

More information about this series at http://www.springer.com/series/8876

Guofeng Zhang

Environmental and Social-economic Impacts of Sewage Sludge Treatment

The Evidence of Beijing

 Springer

Guofeng Zhang
Institute of Economic and Trade
Shijiazhuang University of Economics
Shijiazhuang
China

This research was supported by the Key Discipline of Population, Resources and Environmental Economics of Hebei Province, China and Dr. Scientific Research Foundation, Shijiazhuang University of Economics (BQ201504).

ISSN 2191-5504   ISSN 2191-5512   (electronic)
SpringerBriefs in Economics
ISBN 978-981-287-947-9   ISBN 978-981-287-948-6   (eBook)
DOI 10.1007/978-981-287-948-6

Library of Congress Control Number: 2015954343

Springer Singapore Heidelberg New York Dordrecht London

Springer Science+Business Media Singapore Pte Ltd. is part of Springer Science+Business Media
(www.springer.com)

Acknowledgments

I would like to express my deepest gratitude to my supervisor, Prof. Yoshiro Higano, Doctoral Program in Sustainable Environmental Studies, University of Tsukuba, for his careful guidance, encouragement, and patience during my doctoral study. Without his consistent and illuminating instruction, this thesis could not have reached its present form.

I would like to express my heartfelt gratitude to other members of the Advisory Committee, Prof. Zhenya Zhang and Prof. Helmut Yabar who have instructed and helped me a lot in the past 2 years. Sincere thanks for their comments and advice through the whole process of thesis writing.

I am also greatly indebted to my lab mates at the Higano Laboratory for their valuable suggestions and comments during the study. Thanks from the bottom of my heart go to Prof. Takeshi Mizunoya and Dr. Feng Xu.

I am particularly indebted to Prof. Jinghua Sha and Dr. Jingjing Yan, at China University of Geosciences, Beijng. They give me a lot of support in both studies and daily life. Sincere thanks for their help and encouragement.

Special thanks go to the China Scholarship Council (CSC) and China University of Geosciences, Beijing for awarding me the National Construction High Level Graduate University Government-sponsored Graduate Student Project.

I also owe my sincere gratitude to my friends and my classmates in Higano Lab, Tunyan Wang, Yu Zou, Nan Xiang, Yang Li, Shanshan Wang, Qian Zhou, Keyu Lu, Wei Yang, Junnian Song, Shuai Zhong, Lili Feng, and Shuhong Li, who help me in both studies and daily life. Thanks for their help and for the wonderful time spent with me in University of Tsukuba, we have so many memories.

I am deeply thankful to Prof. Jugang Zhang, Prof. Dezhi Liu, Prof. Jiangao Niu, Prof. Zhenying Yu, and all colleagues in the Department of Economics at Shijiazhuang University of Economics, who give me professional advice for this book.

Lastly, my thanks would go to my beloved family for their loving consideration and great confidence in me all through these years. Thanks to my parents Mr. Cheng Zhang and Mrs. Yahuan Xie, who have always supported my studies.

I owe my loving thanks to my wife Mrs. Xiaojing Ma, who has lost a lot due to my studies abroad. Without her understanding and encouragement, it would have been impossible for me to accomplish my doctoral studies. I am especially obliged to my son Mr. Feiyang Zhang, who is only two and half years old, though he does not know what happened. Every time, when he said I love you in the video, it provided me emotional support.

Contents

List of Figures

List of Tables

Abstract

With the rapid development of urbanization and industrialization process, a sharp increase in urban population and the expansion of urban areas, the volume of industrial wastewater and municipal sewage has increased significantly. In order to improve water quality, the Chinese government has invested heavily in the construction of sewage treatment plants. At the same time, as the by-product of sewage treatment, the amount of sewage sludge has also increased year after year. However, the government has ignored sewage sludge treatment. In terms of current sewage sludge treatment, landfill, aerobic fermentation, natural drying, sludge burning, and without any treatment account for 65 %, 15 %, 3 %, 2 %, and 15 %, respectively. About 80 % of sewage sludge is not treated and disposed in a safe and effective manner. The untreated or improperly treated sewage sludge has polluted the water environment. Beijing is a typical case in this regard in China.

Owing to its rapid economic and population growth, Beijing's municipal sewage emissions are increasing every year. In 2010, the city discharged more than 1.4 billion tons of sewage. Many sewage treatment plants that have been constructed by the Beijing municipal government have adopted advanced technologies, and the sewage treatment rate increased to 80 % in 2010. However, the amount of sewage sludge has also increased. Sewage sludge is the by-product of sewage treatment, and these by-products pollute the water environment. If this sewage sludge cannot be treated properly, 50 % of the water pollutants not removed by sewage treatment will return to the environment. However, the need for sewage sludge treatment has not been addressed by the government. The rate of sewage sludge treatment was less than 50 % in 2010. More than 20,000 tons of total nitrogen is discharged by untreated sewage sludge every year, which accounts for approximately 30 % of the total nitrogen load in Beijing.

Recently, the government has realized the importance of environmental protection. Accordingly, The Twelfth Five-Year Plan of Economic and Social Development requires that all sewage sludge be treated by 2015 and load of chemical oxygen demand (COD) be reduced by 8.7 % in 2015 compared with 2010. Therefore, the government has adopted an integrated policy that promotes water conservation, reduction of working capital, and the introduction of advanced sewage and sewage sludge treatment technologies. In order to determine the optimal development plan for Beijing, it is beneficial to use a simulation method to evaluate the regional, environmental, and economic impacts of adopting advanced technologies for the treatment of sewage sludge.

In this study, a comprehensive optimization simulation model was constructed to simulate socio-economic and environmental development in Beijing city. The simulation period is from 2010 to 2020. The simulation model for this comprehensive analysis was performed using Lingo software.

Based on the comprehensive simulation results, we found that the integrated policy emphasizing on sewage sludge treatment with advanced technology is effective to reduce environmental pollutants and achieve economic development. Specifically, in the optimal scenario (Scenario 4), the total GRP for 2010–2020 reaches 24,151 billion CNY and the average rate of economic growth from 2010 to 2020 is 8.03 %. Moreover, the total net load of T-P, T-N, and COD is 49,000 tons, 449,000 tons, and 1,933 thousand tons, respectively. The reduction rate of T-P, T-N, and COD is 46 %, 40 %, and 25 %, respectively, in 2020 compared with 2010 while keeping the target of 8 % GRP growth.

If the reduction rate of COD is 25 % in 2020 compared with 2010, the reduction rate of energy consumption intensity and GHG emission intensity reach 39 % and 36 %, respectively, both of which can achieve the government's plan. Moreover, GHG emission reduction potential by proper treatment of sewage and sewage sludge is 1.9 million tons for the study period. Additionally the model generates 724,000 TCE as by-product.

The optimal budget expenditures for the policy are 3.19 billion CNY for new sewage plant construction and 6.81 billion CNY for new sewage sludge plant construction for the study period. Specifically, the optimal subsidy for sewage and sewage sludge plants construction of every subregion is: 7,120 million CNY, 484 million CNY, 453 million CNY, 404 million CNY, 429 million CNY, 176 million CNY, 206 million CNY, 245 million CNY, 240 million CNY, 146 million CNY, and 97 million CNY for Central City, Fangshan, Tongzhou, Shunyi, Changping, Daxing, Mentougou, Huairou, Pinggu, Miyun, and Yanqing, respectively.

The optimal sewage and sewage sludge plants construction plan involves 11 new sewage plants featuring the MBR technology and 14 plants featuring the EMBR technology; 9 sewage sludge plants with the A-D-F-I technology, 15 sewage sludge plants with the A-D-F-II technology and 1 sewage sludge plant with the F-C-II technology.

In this study, we also confirm that uniform policies do not always lead to identical subregional impacts. There are regional gaps regarding sustainable development. By comparing the simulation results, Central City (zone 1) plays a key point of sustainable development of Beijing city. The size of GRP will be about 75 % of total GRP of Beijing City in the next 11 years, while water pollutants intensity is lower than the average level of the other subregions.

Chapter 1
Introduction

Abstract Water shortages and water pollution in China is a serious situation that still has not been fundamentally reversed. In order to improve water quality, many sewage plants has been constructed. However, sewage sludgy as the by-product of sewage treatment is ignored by Chinese government. The purpose of our research is to evaluate the environmental and economic impacts with a dynamic comprehensive optimization simulation model, and suggest an optimal policy can achieve the goal of environmentally sustainable economic development. Beijing is selected to make empirical research. The background, objectives, statement of the problem, and literature reviews are shown in this chapter.

Keywords Sewage sludge · Water pollution · Water quality

1.1 Background

Urbanization is a basic process, an important symbol, and an inevitable stage of modernization. According to *National Economic and Social Development Twelfth Five-Year Plan*, urbanization level of China is expected to reach and exceed 50 % in 2015 (The National Development and Reform Commission 2011). With the rapid development of urbanization and industrialization process, a sharp increase of urban population and the expansion of city size and industrial wastewater and municipal sewage are discharging significantly. In 2010, the total amount of sewage was 37.9 billion tons in China. In order to improve water environmental situation, China has increased investment in the construction of a sewage treatment plant. Until 2010, the number of sewage treatment plants is more than 2,500 and the total capacity of sewage treatment is 41.2 billion tons per year. The average rate of sewage treatment is more than 70 % (Ministry of Housing and Urban-Rural Development 2011).

In the same time, as the by-product of sewage treatment, sewage sludge is increasing year after year. The amount of sewage sludge is 30 million tons (80 % moisture) in 2010. However, the government has ignored sewage sludge treatment.

© The Author(s) 2016

G. Zhang, *Environmental and Social-economic Impacts of Sewage Sludge Treatment*, SpringerBriefs in Economics, DOI 10.1007/978-981-287-948-6_1

In terms of the wage of sewage sludge treatment, landfill, aerobic fermentation, natural drying, sludge burning, and without any treatment it accounts for 65, 15, 3, 2, and 15 %, respectively. About 80 % of sewage sludge is not harmless treatment. These untreated and improperly treated sewage sludge has polluted the water environment (Ministry of Environmental Protection 2011). Water environment is still serious. 40.1 % of surface water quality is considered to be less than "IV" class which cannot be used for domestic water and industrial water. The amount of chemical oxygen demand (COD) is 12.38 million tons, in which 35 % is discharged by industrial wastewater and the remaining 65 % is discharged by domestic sewage (Ministry of Environmental Protection 2011).

Beijing is a typical city in point. Owing to its rapid economic and population growth, Beijing's municipal sewage emissions are increasing each year. In 2010, the city produced more than 1.4 billion tons of sewage emissions (Beijing Environmental Protection Bureau 2011). Many sewage treatment plants that have been constructed by the Beijing municipal government have adopted advanced technologies, and the sewage treatment rate increased to 80 % in 2010 (Beijing Water Authority 2011). However, the amount of sewage sludge has also increased. Sewage sludge is the by-product of sewage treatment, and these by-products pollute the water environment. If this sewage sludge cannot be treated properly, 50 % of the water pollutants removed by sewage treatment will return to the environment (Yang 2010). However, the need for sewage sludge treatment has not been addressed by the government. The rate of sewage sludge treatment was less than 50 % in 2010. More than 20,000 tons of total nitrogen is emitted by untreated sewage sludge every year (Zhou 2011), which accounts for approximately 38 % of the total net load of total nitrogen in Beijing (Table 1.1).

In 2010, water shortages and water pollution in urban river downstream serious situation still has not been fundamentally reversed. The river water quality in Beijing is bad compared with other types of water class; 43.2 % of the river water is considered to be "inferior V" class, which is the worst level in China. 6.3 % of lake water is considered to be "inferior V" class (Beijing Environmental Protection Bureau 2010).

Recently, the government has realized the importance of environmental protection. Accordingly, The Twelfth Five-Year Plan of Economic and Social Development (Beijing Municipal Development and Reform Commission 2011) requires that all sewage sludge be treated by 2015 and load of COD be reduced by 8.7 % in 2015 compared with 2010. Therefore, the government has adopted an integrated policy that promotes water conservation, reduction of working capital and the introduction of advanced sewage and sewage sludge treatment technologies.

Therefore, to determine the optimal development plan for Beijing, it is beneficial to use a simulation method to evaluate the regional environmental and economic impacts of adopting advanced technologies for the treatment of sewage sludge.

Table 1.1 Water pollutants discharging in 2010 in Beijing

Water Pollutants	T-P		T-N		COD	
	Amount (tons)	Percentage %	Amount (tons)	Percentage %	Amount (tons)	Percentage %
Discharged after sewage treatment	1,128	24.47	3,598	7.03	85,910	43.08
Discharged due to sewage without treatment	1,862	40.40	16,835	32.90	108,445	54.38
Discharged after sewage sludge treatment	0	0.00	0	0.00	0	0.00
Discharged due to sewage sludge without treatment	1,099	23.84	19,567	38.24	1,668	0.84
Discharged by non-point	520	11.28	11,169	21.83	3,406	1.71
Subtotal	4,609	100	51,169	100	199,429	100
Discharged by rainfall	778	–	7,379	–	3,432	–
Total	5,387	–	58,548	–	202,861	–

1.2   Statement of the Problem

In this study, we select Beijing as the study area, and construct a comprehensive simulation model to evaluate the economic and environmental impact of sewage sludge treatment by using advanced technology. The comprehensive simulation model includes socio-economic activities, water pollutants reduction, energy consumption and supply and greenhouse gas (GHG) emission. Based on the simulation result an integrated policy will be proposed to improve environmental situation without deterioration of economic development in Beijing.

Based on the current situation analysis in the study area, we make a study route as follows:

Step 1, we construct a simple original simulation model which includes economic model and water pollutant model to verify the feasibility of the environment policies. Step 2, if the initial simulation results indicate that it is feasible to treat sewage sludge as a crucial factor to achieve local government targets, we will construct a comprehensive simulation model which includes economic mode, water pollutant model, GHG model and energy model. But if the original simulation result is infeasibility to achieve government target, we repeat step 1 and propose new policies. Step 3, based on the comprehensive simulation results, an integrated policy will be proposed to achieve socio-economic and environmental sustainable development in Beijing.

The research frame work is introduced in Fig. 1.1.

There are five chapters in this dissertation. This chapter introduces research background, content, objective and literature review. Chapter 2 is descriptive

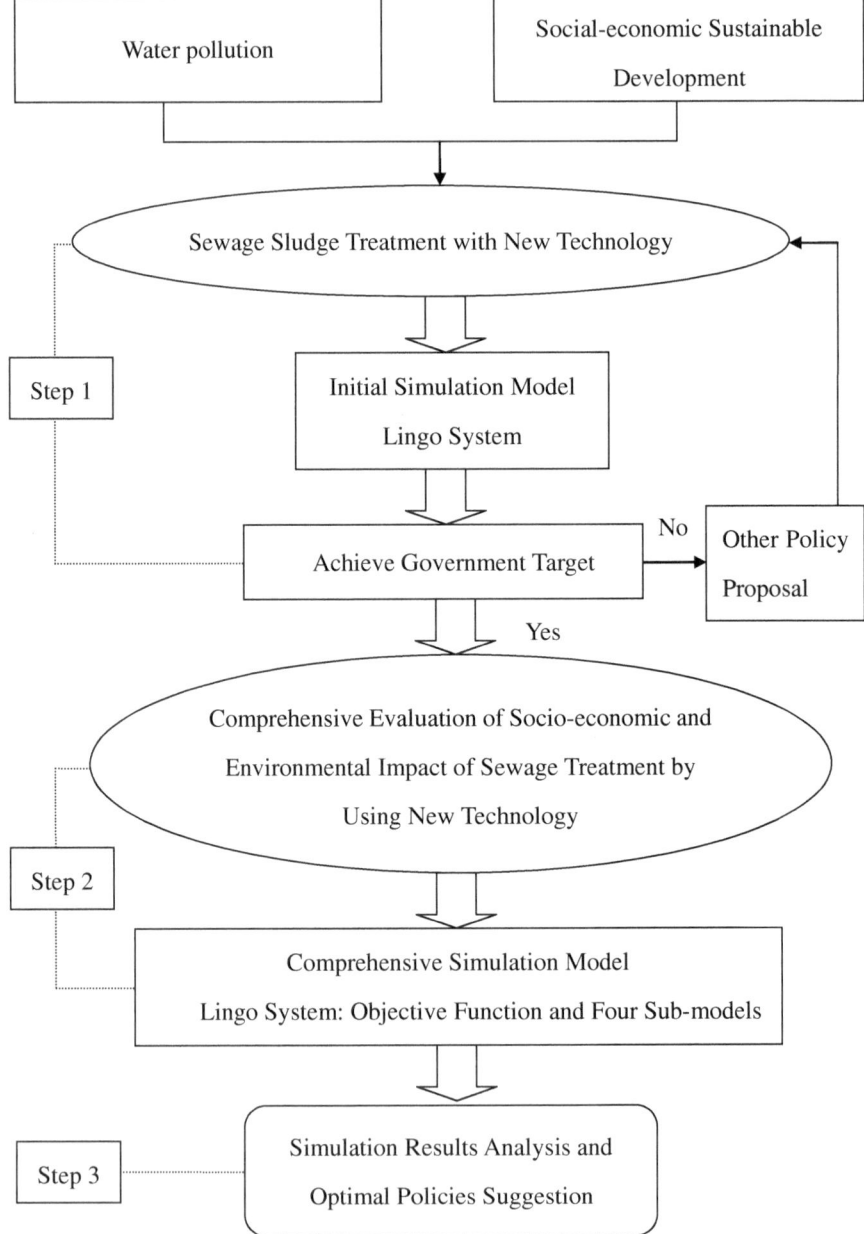

Fig. 1.1 Research framework

analysis of the present socio-economic and environmental condition in Beijing include GRP, population, financial income and expenses, water environmental condition, energy consumption and GHG emission. Chapter 3 represents the concept of optimal environmental policies and reconstructs the integrated comprehensive simulation model with introducing various advanced sewage and sewage sludge treatment technologies. Chapter 4 analyzes the comprehensive simulation results and puts forward optimal policies. Chapter 5 concludes the thesis by summarizing the major finding and further research ideas.

1.3 Objective of the Study

In light of the above discussion, the purposes of this paper are as follows:

(i) Establish a dynamic comprehensive optimization simulation model which considers water pollutant emission, energy consumption, GHG emission and gross regional product (GRP) growth and simulate this model via Lingo software which is a fast, efficient and straightforward tool for solving linear and nonlinear optimization problems (Li and Higano 2007).

(ii) Evaluate the water pollutants reduction, energy consumption intensity, GHG emission intensity and rate of GRP growth impacts of sewage sludge treatment with using advanced technologies.

(iii) Suggest an optimal policy can achieve the goal of environmentally sustainable economic development of Beijing.

1.4 Literature Review

Recently, China is in urgent need of effective policies to reduce water pollutants and to improve water environment. There are a lot of researches on water environment improvement policy evaluation. In this part, the literature review is summarized in five aspects: research on water conservation policy evaluation, research on reduction working capital policy evaluation, research on membrane bioreactors (MBRs) water treatment technology evaluation, research on sewage sludge treatment evaluation, research on comprehensive evaluation model based on input–output table.

1.4.1 Research on Water Conservation Policy Evaluation

In the view of research content, previous studies on water conservation policy evaluation focus on establishing a comprehensive evaluation system. However, these studies only ex-post evaluate the impact of policy, but not predict. In term of research method, most of these studies use analytic hierarchy process

(AHP) analysis and expert surveys, questionnaires and so on. These methods have subjective factors in determining the index weight and score. The different policy evaluation results may be caused by different survey and different experts. Specific research is as follows.

Wang (2003) comprehensively evaluated the effect of Small watershed management of the Loess Plateau in Shanxi province by using AHP and multilevel fuzzy comprehensive evaluation method. He considered that it is necessary to develop ecological agriculture in order to reduce management cost and to improve management efficiency of soil erosion. Chen (2007) focused on the policy benefit evaluation index system and water conservation and established a regional soil and water conservation benefit evaluation index system based on the basis of efficiency, ecological, economic and social benefits. Bao (2008) evaluated the soil conservation policy by using expert survey and focused on the problems in the process of policy implementation. He believed that departmental coordination, management practices, and legal constraints are the main reason that affecting the implementation of soil and water conservation policies. Bao (2008) analyzed the main influencing factors of soil and water conservation investments and focused on construction of evaluation framework. In order to determine the external effects, he suggested that soil erosion and water conservation evaluation should include not only the costs and benefits of individual investment analysis, but also impact on other industries based on the socio-economic impact analysis. Yao (2009) evaluated the implementation of soil and water conservation policies in Loess Plateau region by using of questionnaires. He constructed a comprehensive evaluation index system in which the economic benefit is the most important indicator of evaluation. Wang (2010) established a comprehensive evaluation system based on economic, social, and ecological benefits evaluation and made a comprehensive evaluation of small watershed soil erosion in Loess Hilly Area by using the AHP method.

1.4.2 Research on Reduction Working Capital Policy Evaluation

Many studies on reduction working capital policy evaluation have focused on the environmental impact of industrial restructuring, environmental impact coefficient of industry sectors and establishment of evaluation index system. These studies usually used econometric models to ex-post evaluation of the polity impact of the study area, and lacked of policy impact predictions. Specific research is as follows.

Cui and Yang (2010) studied the environmental impact of industrial restructuring and proposed quantitative expression the relationship of human social and economic activities, resources and environment by using environmental carrying capacity. He constructed a comprehensive evaluation index system based on economic benefit, environmental resource productivity, and environmental resources satisfaction. Li (2002) analyzed the changes in energy consumption and output of industrial sector in Tangshan City. He believes that industrial restructuring is the fundamental way of economic development and environmental protection. Peng (2005) calculated the ecological impact coefficient of different types of industries and the regional industrial structure and quantitatively evaluated the ecological environmental impact of industrial restructuring from 1992 to 2003 in the Lijiang City of Yunnan province. He proposed that in order to reduce resource depletion and environmental pollution, government should restrict high-polluting industries and develop circular economy industrial. Hu (2008) analyzed the three major factors restricting the economic and the environmental sustainable development of China. He considered unreasonable industrial structure is the primary reason of the ecological environmental deterioration and proposed the establishment of a strategic environmental planning and evaluation system which is suitable for China. Wang (2008) evaluated the wastewater discharging and GHG emissions impact of the industrial structure adjustment in Jiangsu province. He suggested that in order to improve environmental situation, government should restructure industrial sector. Zhang (2008) evaluated the regional ecological environmental impact of different industry by using the fuzzy comprehensive evaluation method, in Chuzhou City of Anhui province and calculated the ecological impact coefficient of different types of industries and the regional industrial. Zhao (2008) constructed an econometric model of the ecological effects of industrial restructuring, and quantitatively evaluated the ecological environmental impact of industrial structure adjustment, from 1970 to 2006 in Jiangsu province. He believed that industrial restructuring can improve the ecological environment, but improvement trend is more and more slowly, new policy introducing is necessary. Wang (2013) studied how to use resources and environmental efficiency analysis method to evaluate the environmental impact of industrial restructuring and introduced the concept of method, procedures, application examples, precautions, and limitations.

1.4.3   Research on Membrane Bioreactors (MBRs) Water Treatment Technology Evaluation

Advanced technologies for sewage treatment are often advocated in order to decrease the impact of water environmental pollution. Membrane bioreactors (MBRs) are now widely used for municipal and industrial wastewater treatment because of its compact plant footprint and high-quality effluent (Gunder and Krauth 1998; Lin et al. 2011; Kimura et al. 2005; Krauth and Staab 1993; Zhang et al. 2003; Steven et al. 1998; Lv et al. 2006; Zheng et al. 2005). However, several disadvantages, such as the high operation cost, high energy consumption and membrane fouling, limit its future utilization (Chriemchaisri et al. 1993; Trouve et al. 1994; Yoona et al. 2004). It is necessary to apply a comprehensive evaluation of environmental and economic impact of MBRs, whereas a few studies have been conducted to comprehensively evaluate with respect to environment and socio-economy.

In the view of research content, previous studies on the MBRs technology are roughly divided into four portions: (1) membrane fouling, (2) effluent quality, (3) energy consumption, and (4) cost considerations (Nieuwenhuijzen et al. 2008). The former two studies mainly focus on the optimal method to release the membrane fouling (Meng et al. 2009; Nywenind and Zhou 2009; Kimura et al. 2005; Le-Clech et al. 2006) and the efficiency of water pollutants reduction (Lesjean et al. 2002; Laera et al. 2012; Kimura et al. 2005; Liang et al. 2010), respectively, while the latter two intend on the assessment of energy consumption and the operation/maintenance of the MBRs technology (Owen et al. 1995; Fletcher et al. 2007; Lin et al. 2011; Gander et al. 2000; Côté et al. 2004; Le-Clech et al. 2006; Churchouse et al. 1997; Yoona et al. 2004; Zhang et al. 2003). By contrast, the comprehensive evaluation of environmental and socio-economic impacts was rarely investigated.

The researches of MBRs technology assessment can be divided into two categories by the varied methods, i.e., comparison method and life cycle analysis (LCA). For comparison method, the investment scale and the costs, including running cost, maintenance cost, and power consumption, are analyzed to find out the cheapest option from different mixed procedures of MBRs (Owen et al. 1995; Fletcher et al. 2007; Gander et al. 2000; Côté et al. 2004; Le-Clech et al. 2006; Churchouse et al. 1997; Yoona et al. 2004; Zhang et al. 2003). The weakness of this kind of researches lies on the less consideration of the technology on the environment or socio-economy. Comparatively, the studies of LCA assessed MBRs from environmental and cost aspects rather than the impacts on socio-economy (Lin et al. 2011; Tangsubkul et al. 2005).

1.4.4 Research on Sewage Sludge Treatment Assessment

Many studies have estimated the impacts of sewage sludge treatment on economic development and the environment. While some of these studies have performed economic evaluations, which focused on analyses of capital investment and operation cost of sewage sludge treatment (Kim and Parker 2008; Shi 2009); those studies, however, ignore the environmental impacts of sewage sludge treatment. A life cycle assessment is one method that can be used to perform a comprehensive evaluation of the technical, economic, and environmental aspects of sludge treatment (Murray et al. 2008; Hong et al. 2009; Enrica et al. 2011). However, this method cannot select the optimal sewage sludge treatment technology. The AHP method has been used in China to evaluate the technological and economic impacts of sewage sludge treatment (Mao et al. 2010), but the use of expert scoring in this approach is subjective, as different experts may provide differing evaluations.

1.4.5   Research on Comprehensive Evaluation Model Based on Input–Output Table

From the beginning of the 1960s, the input–output model is extended to the field of environmental studies and natural resources. Ayre and Kneese (1969), Leontief (1977), Lee (1982), Forsund (1985), Perrings (1987) focused on the relationship of resources and pollutants between different industries. Some studies focused on how to use the input–output table to establish a comprehensive evaluation model. Turner et al. (2000) proposed that construct a comprehensive system model which is based on input–output table to predict the changes of investment, industry production, pollutants of industry sector and other economic activity in different cases. Hannon (2001) believes that establishing such a comprehensive evaluation model based on input–output table, natural resources and environmental substances using physical input unit, and the economic development variables using monetary unit. Moffatt and Hanley (2001) believes that evaluation of the impact on economy by different environmental polices is an important function of this comprehensive evaluation model. And it must consider many factors, such as the impact of environmental policies on the total output, change of industrial structural, and the long-term economic growth. In a sum, the comprehensive model based on input–output table is a popular tool for environmental policy (Oloveira 2004; Wei 2006) and evaluation empirical studies have been carried out in many countries and regions. Some of these studies focused on the impact on GHG emission intensity by energy consumption policy, such as in Australia (Lenzen 1998), in Portugal (Oliveira 2004), in Korea (Chung 2009), in Greece (Hristu-Varsakelis 2010), in China (Yan 2010). The other study focused on the water pollutants reduction policy evaluation. Higano and Sawada (1997), Higano and Yoneta (1999), Hirose and Higano (2000), Mizunoya (2007) established a comprehensive water environment improving policy evaluation model and selected the second largest lake in Japan (Kasumigaura Lake) watershed as the basis for empirical research by using computer simulation. Some Chinese scholars learned and developed this model, and established environmental policy evaluation model in China. Xu (2009) comprehensively evaluated the water environment policy of Qinhuangdao Yanghe Reservoir. Yan (2010) established suburbs livestock industry biomass utilization evaluation model and selected Miyun County of Beijing City as an example of empirical research. Li (2012) established a comprehensive recycling of electronic waste environmental policy evaluation model. Xiang (2012) constructed a comprehensive social and economic and environmental evaluation model and evaluated reclaimed water utilization policy in Tianjin City.

Summary, the integrated approach has been used in modeling since the late 1960s when the input–output model has been already applied in environmental and natural resource field. The optimization and comprehensive evaluation model based on material balance, energy balance, and economic balance is suitable for evaluating water treatment technologies. Many studies using this approach have been done to assess wastewater treatment technologies, but in these studies the energy

Table 1.2 Literature review

Research	Authors	Year	Achievements	Limitations
Research on water conservation policy evaluation	Wang	2003	Evaluated the effect of small watershed management of the Loess Plateau in Shanxi province	These studies only ex-post evaluated the impact of policy, but did not predict
	Chen	2007	Established a regional soil and water conservation benefit evaluation index system	
	Bao	2008	Evaluated the soil conservation policy by using expert survey	
	Yao	2009	Evaluated the implementation of soil and water conservation policies in Loess Plateau region	
	Wang	2010	Made a comprehensive evaluation of small watershed soil erosion in Loess Hilly Area by using the AHP method	
Research on reduction working capital policy evaluation	Cui	1998	Constructed an index system based on economic benefit, environmental resource productivity, and environmental resources satisfaction	These studies focused on environmental impact coefficient of industry sectors and establishment of evaluation index system, lacked of policy impact predictions
	Li	2002	Analyzed the changes in energy consumption and output of industrial sector in Tangshan City	
	Peng	2005	Calculated the ecological impact coefficient of different types of industries and the regional industrial structure	
	Hu	2008	Analyzed the three major factors restricting the economic and the environmental sustainable development of China	
	Wang	2008	Evaluated the wastewater discharging and GHG emissions impact of the industrial structure adjustment	

(continued)

Table 1.2 (continued)

Research	Authors	Year	Achievements	Limitations
Research on reduction working capital policy evaluation	Zhang	2008	Evaluated the regional ecological environmental impact of different industries	These studies focused on environmental impact coefficient of industry sectors and establishment of evaluation index system, and lacked policy impact predictions
	Zhao	2008	Constructed an econometric model of the ecological effects of industrial restructuring	
	Gao	2011	Estimated the regional environmental pollution coefficient of industrial restructuring	
	Wang	2013	Use resources and environmental efficiency analysis method to evaluate the environmental impact of industrial restructuring	
Research on Membrane bioreactors (MBRs) water treatment technology evaluation	Owen	1995	Evaluated investment, operating costs, and energy consumption of MBR technology, by using of technical and economic evaluation method and found the most economical combination of MBR process	The weakness of this kind of researches lies on the less consideration of the technology on the environment or socio-economy
	Churchouse	1997		
	Gander	2002		
	Zhang	2003		
	Côté	2004		
	Fletcher	2007		
	Tangsubkul	2005	Comprehensively evaluated the impact of running costs and energy consumption of MBR technology	Did not consider the impact on other sectors of economic development
	Lin	2011		
Research on sewage sludge treatment assessment	Kim	2008	Focused on analyses of capital investment and operation cost of sewage sludge treatment	Ignore the environmental impacts of sewage sludge treatment
	Shi	2009		
	Mao	2010	Evaluate the technological and economic impacts of sewage sludge treatment by AHP method	The use of expert scoring in this approach is subjective
	Murray	2008	Comprehensively evaluated the technical, economic and environmental aspects of	This method cannot select the optimal sewage sludge treatment technology
	Hong	2009		
	Enrica	2010		

(continued)

Table 1.2 (continued)

Research	Authors	Year	Achievements	Limitations
			sludge treatment by LCA method	
Research on comprehensive evaluation model based on input–output table	Ayre	1969	Theoretical study on the use of input–output tables to establish a comprehensive evaluation model	Lack of empirical research and prediction
	Leontief	1977		
	Lee	1982		
	Forsund	1985		
	Perrings	1987		
	Turner	2000		
	Moffatt	2001		
	Hannon	2011		
	Lenzen	1998	Focused on the impact on GHG emission intensity by energy consumption policy	Lack of comprehensive policy evaluation
	Oliveira	2004		
	Wei	2006		
	Chung	2009		
	Hristu	2010		
Research on comprehensive evaluation model based on input–output table	Higano	1997	Established a comprehensive water environment improving policy evaluation model	Did not consider energy consumption and GHG emission in this model
		1999		
	Hirose	2000		
	Mizunoya	2007		
	Xu	2009		
	Xiang	2012		
	Yan	2010	Established a comprehensive evaluation model which included both water policy and energy policy	Did not consider zone model in the submodel of energy balance
	Li	2012	Established a comprehensive recycling of electronic waste environmental policy evaluation model	Did not have zone model in this study
	Wang	2013	Established a comprehensive environmental policy evaluation model and focused on energy consumption and GHG emission	

consumption and greenhouse gas (GHG) emission have been little taken into account. The compare of achievement and limitations of all the previous studies is in Tables 1.1 and 1.2.

References

Ayres RU, Kneese AV (1969) Production, consumption and externalities. Am Econ Rev 59 (3):282–297

Bao XB (2008) The impact assessment of soil and water conservation investment. J Econ Water Resour 26(4):23–26

Beijing Environmental Protection Bureau (2011) Beijing Environment Bulletin 2010 [EB/OL]. http://www.bjepb.gov.cn/portal0/tab181/, 3 Jun 2011 (in Chinese)

Beijing Municipal Development and Reform Commission (2011) The Twelfth Five-Year Plan for the National Economic and Social Development of Beijing [EB/OL]. http://www.bjpc.gov.cn/fzgh_1/guihua/12_5/Picture_12_F_Y_P/, 21 Jan 2011

Beijing Water Authority (2011) Beijing Water Resources Bulletin 2010 [EB/OL]. http://www.bjwater.gov.cn/tabid/207/Default.aspx, 8 Mar 2011 (in Chinese)

Chen QC, Zhang RS (2007) Study on approaches and index system for quantitatively analyzing comprehensive benefits of soil and water conservation. J China Inst Water Resour Hydropower Res 5(2):95–104

Chriemchaisri C, Yamamoto Y, Vigneswaran S (1993) Household membrane bioreactor in domestic wastewater treatment. Water Sci Technol 27:171–178

Chung W, Tohno S, Shim S (2009) An estimation of energy and GHG emission intensity caused by energy consumption in Korea: an energy IO approach. Appl Energy 86(10):1902–1941

Churchouse S (1997) Membrane bioreactors for wastewater treatment-operating experiences with the Kubota submerged membrane activated sludge process. Membr Technol 1997(83):5–9

Côté P, Masini M, Mourato D (2004) Comparison of membrane options for water reuse and reclamation. Desalination 167(15):1–11

Cui FJ, Yang YS (2010) The assessment on the influence of industry structure on urban ecological environment. China Environmental Science 10(1):37–42 (in Chinese).

Enrica U, Ivet F, Jordi M, Joan G (2011) Technical, economic and environmental assessment of sludge treatment wetlands. Water Res 45:73–582

Fletcher H, Mackley T, Judd S (2007) The cost of a package plant membrane bioreactor. Water Res 41(12):2627–2635

Forsund FR (1985) input–output models: national economic models and the environment. In: Kneese AV, Sweeney JL (eds) Handbook of national resource and energy economics, vol 1 p 6

Gander M, Jefferson B, Judd S (2000) Aerobic MBRs for domestic wastewater treatment: a review with cost considerations. Sep Purif Technol 18:119–130

Gander M, Jefferson B, Judd S (2002) Aerobic MBRs for domestic wastewater treatment: a review with cost considerations. Sep Purif Technol 18:119-130

Gao ZG, Zhao XW (2011) An empirical research on the relationship between regional industrial structure adjustment and Environment Pollution in Xinjiang Based on the Panel Data Model. Ecol Econ 1:82–95

Gunder B, Krauth K (1998) Replacement of secondary clarification by membrane separation-results with plate and hollow fibre modules. Water Sci Technol 38:383–393

Hannon B (2011) Ecological pricing and economic efficiency. Ecol Econ 36:19–30

Higano Y, Sawada T (1997) The dynamic policy to improve the water quality of lake Kasumigaura. Stud Reg Sci 26(1):75–86

Higano Y, Yoneta A (1999) Economic policies to relieve contamination of Lake Kasumigaura. Stud Reg Sci 29(3):205–218

Hirose F, Higano Y (2000) A simulation analysis to reduce pollutants from the catchment area of lake Kasumigaura. Stud Reg Sci 30(1):47–63

Hong JL, Hong JM, Otaki M, Jolleit O (2009) Environmental and economic life cycle assessment for sewage sludge treatment processes in Japan. Waste Manag 29:696–703

Hristu VD, Karagianni S, Pempetzoglouc M, Sfetsos A (2010) Optimizing product with energy and GHG emission constraints in Greece: an input–output analysis. Energy Policy 38:1566–1577

Hu LC, Zhang SC (2008) Analysis on content and methods of environmental impact evaluation of industrial structure adjustment. Environ Prot 406:16–19

Kim Y, Parker W (2008) A technical and economic evaluation of the pyrolysis of sewage sludge for the production of bio-oil. Bioresour Technol 99:1409–1416

Kimura K, Yamato N, Yamamura H, Watanabe Y (2005) Membrane fouling in pilot-scale membrane bioreactors (MBRs) treating municipal wastewater. Environ Sci Technol 39:6293–6299

Krauth K, Staab KF (1993) Pressurized bioreactor with membrane separation for wastewater treatment. Water Res 27:405–411

Laera G, Cassano D, Lopez A, Pinto A, Pollice A, Ricco G, Mascolo G (2012) Removal of organics and degradation products from industrial wastewater by a membrane bioreactor integrated with ozone or UV/H2O2 treatment. Environ Sci Technol 46:1010–1018

Le-Clech P, Chen V, Fane TAG (2006) Fouling in membrane bioreactors used in wastewater treatment. J Membr Sci 284(1–2):17–53

Lee KS (1982) A generalized input–output model of an economy with environmental protection. Rev Econ Stat 64(3):466–473

Lenzen M (1998) Primary energy and greenhouse gases embodied in Australian final consumption: an input–output analysis. Energy Policy 26(6):495–506

Leontief WW (1977) The future of the world economy. Oxford University Press, New York

Lesjean B, Gnirssb R, Adamc C (2002) Process configurations adapted to membrane bioreactors for enhanced biological phosphorous and nitrogen removal. Desalination 149:217–224

Li B, Higano Y (2007) An environmental socioeconomic framework model for adapting to climate change in China. In: Cooper R, Donaghy K, Hewings G (eds) Globalization and regional economic modeling in advances in spatial science. Springer, New York, pp 327–349

Li WJ, Yang MC, Shi PJ (2002) Analysis for the relationship between Tangshan industrial structure and its environment impacts. Geogr Res 21(4):511–518

Li Y, Zou SX, Xu F, Yabar H, Higano Y (2012) The impact of introducing recycling taxes on China's environmental policy: case study on E-waste recycling. J Sustain Dev 5(4):83–98

Liang Z, Das A, Beerman D, Hu Z (2010) Biomass characteristics of two types of submerged membrane bioreactors for nitrogen removal from wastewater. Water Res 44(11):3313–3320

Lin HJ, Chen JR, Wang FY, Ding LX, Hong HC (2011) Feasibility evaluation of submerged anaerobic membrane bio-reactor for municipal secondary wastewater treatment. Desalination 280:120–126

Lv WZ, Zheng X, Yang M et al (2006) Virus removal performance and mechanism of a submerged membrane bioreactor. Process Biochem 41:299–304

Mao H, Xu DQ, Wang WJ (2010) The evaluation model and application research of sludge disposal method in sewage plant. Environ Sci Manag 35(1):191–184

Meng F, Chae SR, Drews A, Kraume M, Shin HS, Yang F (2009) Recent advances in membrane bioreactors (MBRs): membrane fouling and membrane material. Water Res 43(6):1489–1512

Ministry of Environmental Protection (2011) China Environmental Protection Bulletin 2010. [EB/OL]: http://jcs.mep.gov.cn/hjzl/zkgb/, 3 Jun 2011 (in Chinese)

Ministry of Housing and Urban-Rural Development (2011) China Urban Construction Statistics 2010. Yearbook China Planning Press, Beijing

Mizunoya T, Sakurai K, Kobayashi S, Piao SH, Higano Y (2007) A simulation analysis of synthetic environment policy: effective utilization of biomass resources and reduction of environmental burdens in Kasumigaura basin. Stud Reg Sci 36(2):355–374

Moffatt I, Hanley N (2001) Modeling sustainable development: systems dynamic and input–output approaches. Environ Model Softw 16:545–557

Murray A, Horvath A, Nelson K (2008) Hybrid life-cycle environmental and cost inventory of sewage sludge treatment and end-use scenarios a case study from China. Environ Sci Technol 42(9):163–3169

Nieuwenhuijzen AFV, Evenblij H, Uijterlinde CA, Schulting FL (2008) Review on the state of science on membrane bioreactors for municipal wastewater treatment. Water Sci Technol 57 (7):979–986

Nywenind JP, Zhou H (2009) Inganfluence of filtration conditions on membrane fouling and scouring aeration effectiveness in submerged membrane bioreactors to treat municipal wastewater. Water Res 43(14):3548–3558

Oliveira C, Antunes H (2004) A multiple objective model to deal with economy-energy-environment interactions. Eur J Oper Res 153(2):370–385

Owen G, Bandi M, Howell JA, Churchouse SJ (1995) Economic assessment of membrane processes for water and waste water treatment. J Membr Sci 102:77–91

Peng J, Wang YL, Ye MT, Chang Q (2005) Research on the change of regional industrial structure and its eco-environmental effect: a case study in Lijiang City. Yunnan Province. Acta Geographica Sinica 60(5):798–806 (in Chinese)

Perrings C (1987) Economy and environment. Cambridge University Press, New York

Shi J (2009) Cost-effectiveness analysis and evaluation on the municipal wastewater sludge treatment and disposal system (I). Water Wastewater Eng 35(8):32–35 (in Chinese)

Tangsubkul N, Beavis P, Moore SJ, Lundie S, Waite TD (2005) Life cycle assessment of water recycling technology. Water Resour Manag 19(5):521–537

The National Development and Reform Commission (2011) National Economic and Social Development Twelfth Five-Year Plan. People's Publishing House, Beijing

Trouve E, Urbain V, Manem J (1994) Treatment of municipal wastewater by a membrane bioreactor: results of a semi-industrial pilot-scale study. Water Sci Technol 30:151–157

Turner RK (2000) Integrating natural and socio-economic science in coastal management. J Mar Syst 25:447–460

Wang L (2008) The environmental impact of industrial structure adjustment analysis in Jiangsu province. Mark Wkly Disquis Ed 7:49–50 (in Chinese)

Wang YL (2010) Benefit evaluation of soil erosion control of small watershed in loess hilly and gully regions-taking gaoquan small watershed in Dingxi of Gansu as an example. J Northwest A&F Univ (Social Science Edition) 10(1):37–42

Wang JQ, Chen CG, Li TS (2003) Benefit evaluation of watershed management on Loess Plateau of Shanxi Province. Bull Soil Water Conserv 23(6):61–64 (in Chinese)

Wang XY, Yang LF, Liu R (2013) Application of resource/environment efficiency analysis in environmental impact assessment of regional industry development planning. Environ Sci Manag 38(7):10–12 (in Chinese)

Wei Y (2006) A scenario analysis of energy requirements and energy intensity for China's rapidly developing society in the year 2020. Technol Forecast Soc Chang 73(4):405–421

Xiang N, Xu F, Sha JH, Helmut Y, Higano Y (2012) Comprehensive evaluation of effectively utilizing reclaimed water to accomplish sustainable development in Tianjin. China. Adv Mater Res 524–527:3040–3045

Xu F, Yoshiro H (2009) Comprehensive evaluation of water quality improvement policy in reservoir of Yang He in Qin Huang Dao He Bei China. Stud Reg Sci 39(3):751–766

Yan JJ, Xao RG, Sha JH (2010) Comprehensive evaluation of integrated pollutant-minimization policies in rural areas of China. China Popul Res Environ 20(3):124–127

Yang XP (2010) Status and ideas of the Beijing Municipal sludge disposal. Water Wastewater Inf 8:17–18 (in Chinese)

Yao WB, Liu WZ, Zhao AC, Li HY (2009) Evaluation index of the soil and water conservation benefit. Sci Soil Water Conserv 7(1):112–117

Yoona SH, Kimb HS, Yeomb IT (2004) The optimum operational condition of membrane bioreactor (MBR): cost estimation of aeration and sludge treatment. Water Res 38(1):37–46

Zhang J (2008) Evaluation and regulation of ecological environmental impact of the Regional Industrial Structure. Acta Agriculturae Jiangxi 20(2):129–133 (in Chinese)

Zhang SY, Houten RV, Eikelboom DH, Doddema H, Jiang ZC, Fan YB, Wang JS (2003) Sewage treatment by a low energy membrane bioreactor. Bioresour Technol 90(2):185–192

Zheng X, Lv WZ, Yang M et al (2005) Evaluation of virus removal in MBR using coliphages T4. Chin Sci Bull 50:862–867

Zhao T, Ding P (2008) An empirical analysis of regional industrial structural change impacts on the ecological environment—in Jiangsu Province. Ind Technol Econ 27(12):90–93

Zhou J (2011) The Sludge status and the treatment strategy of Beijing. Water Ind Mark 4:30–32 (in Chinese)

Chapter 2
Current Situation in the Study Area

Abstract Beijing is the capital of the People's Republic of China. It is a typical city in which huge amount of sewage sludge discharged without treatment. Water shortage and water pollution has become key factors which limits Beijing socio-economic and environmental sustainability. In this chapter, the economic and environmental characteristics and the capacity of sewage and sewage sludge treatment in Beijing are presented. Based on the analysis of Beijing situation, it is necessary using a comprehensive simulation method to evaluate the environmental and economic impact of sewage sludge treatment.

Keywords Beijing · Economic development · Water environment · Sewage sludge treatment capacity

Beijing, the capital of the People's Republic of China, is the center of the nation's politics, culture, and international exchanges and a modern metropolis full of vitality.

2.1 History and Relics

Beijing is home to the well-known "Peking Man (Sinanthropus pekinensis)" relic that dated back 200,000–700,000 years ago. Evidence of historical record and excavated relics has proved that the city has stood on its current site for well over 3,000 years. It was the capital city for Liao, Jin, Yuan, Ming, and Qing dynasties in ancient and contemporary Chinese history.

With the founding of People's Republic of China on October 1, 1949, Beijing became the capital of the new republic and has since developed itself into a political and cultural center of China and international exchange hub. Cultural relics in Beijing like the Forbidden City, the Great Wall, Zhoukoudian Peking Man Relics, the Temple of Heaven, the Summer Palace, and the Ming Tombs are world cultural and natural heritage approved by the United Nations. There are a total of 7,309 historical and relic sites in Beijing among which 60 are classified as national cultural heritages and another 234 are Beijing cultural heritages.

G. Zhang, *Environmental and Social-economic Impacts of Sewage Sludge Treatment*, SpringerBriefs in Economics, DOI 10.1007/978-981-287-948-6_2

2.2 Natural Geography

Beijing Municipality is located between 116°20′ east longitude and 39°56′ north latitude, with Tianjin City on its eastern border and Hebei province on the other three sides.

Lying in the northern part of North China Plain, Beijing is surrounded by mountains on the west, north and northeast. The northeastern part of the city is high while the southwestern part is low topographically, with a southeastern plain tilted gradually downward to the Bohai Sea. Major rivers flowing through Beijing include Yongding River, Chaobai River, North Canal, and Juma River that mostly originated in the northeastern mountainous areas of Beijing. These rivers flow through rugged mountains towards southeastern plain of Beijing and in the end join the Bohai Sea.

Beijing has a continental monsoon climate with four distinct seasons. Spring and autumn are short while winter and summer are long. The average temperature of Beijing in year 2002 was 13.2 °C. December was the coldest month with an average temperature of –2.9 °C while July was the hottest month with an average temperature of 27.5 °C. Average annual rainfall in 2002 equaled 370.4 mm, which was relatively low. The frost-free period of Beijing for 2002 was 199 days.

Beijing occupies a total area of 16,807.8 km^2. Around 10,417.5 km^2 are mountainous areas, which constitute 62 % of the total space of Beijing. The plain area of Beijing covers 6,390.3 km^2, which accounts for 38 % of the city's total area. Beijing municipality has 14 subordinate districts and 2 counties.

2.3 Population and Nationalities

By the end of 2010, Beijing has a total of 19.61 million residents among which 12.58 million people are registered permanent ones. Over the past decade (2001–2010), the average annual population growth rate is 3.8 % (Fig. 2.1). The population density of Beijing is 1,166 people per square kilometer.

According to the Sixth Census conducted in year 2010, average life expectancy of Beijing residents reached 80.2 years. Beijing boasts all 56 ethnic nationalities of the country, with nationality of Hui, Manchu, Mongolia, and Korea exceeding 10,000 people.

In term of population distribution, the areas of Central City and other districts have similar size, but the population of the Central City accounted for about 60 % of the total population in Beijing. The population of every subregion in 2010 is shown in Table 2.1.

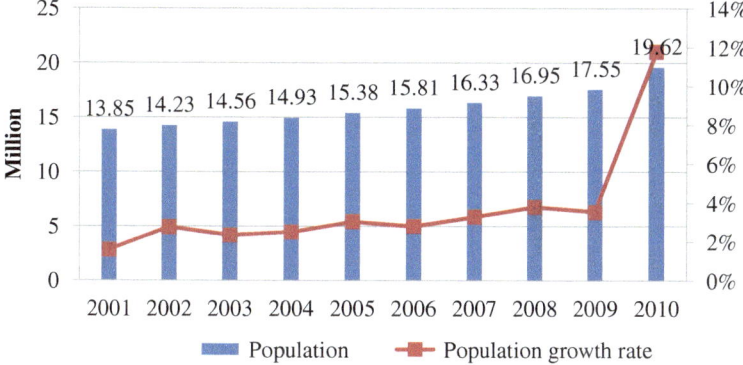

Fig. 2.1 Population growth for Beijing

Subregion	Population	Percentage (%)
Central City	11,716	60
Fangshan	945	5
Tongzhou	1,184	6
Shunyi	877	4
Changping	1,661	8
Daxing	1,365	7
Mentougou	290	1
Huairou	373	2
Pinggu	416	2
Miyun	468	2
Yanqing	317	2
Total	19,612	100

Table 2.1 Population of each subregion in thousands

2.4 Economic Development

Through more than 50 years of construction, Beijing has changed from a consumer city to a major city with various industries. Beijing ranks second among the top 50 cities in China in terms of comprehensive power, and is the first among the 40 best cities in China in terms of investment environment.

As a dynamic city, Beijing has a wide range of industries. During the years after China adopted reform and open-up policies and especially in the Eleventh Five-year period (year 2005–2010), CPC Beijing Committee and Beijing Municipal Government has made adjustments to the city's economic structure and layout to

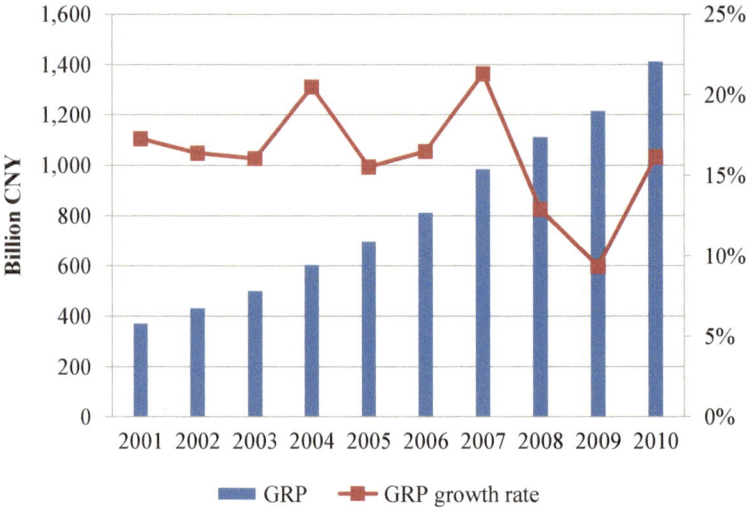

Fig. 2.2 GRP growth trend from 2001 to 2010 for Beijing

ensure a healthy, sustained economic development for the city. In 2010, Beijing's gross domestic production (GDP) increased 16 % from previous year to reach 1,411 billion CNY (Fig. 2.2).

In the year of 2010, Beijing saw the establishment of a modern agriculture structure highlighting cash crops and fruit tree plantation, livestock breeding and eco-tourism. Livestock breeding occupies 55 % of the gross agricultural output. Cash crops accounted for 45.4 % of the total crops. Beijing has a fully integrated industrial structure. It has to date nearly 31,547 industrial enterprises, covering fields of electronics, machinery, chemicals, light industry, and textile and car manufacturing. High tech and modern manufacturing industries have become the leading forces of Beijing's industrial growth. Total value added of the industrial sector of Beijing achieved 339 billion Yuan in 2010. Beijing is also a major city with the greatest development of the tertiary industry with 75 % of the total GDP devoting to the tertiary industry in 2010. Over the past decade, the proportion of three industries structure shown in Fig. 2.3.

The regional economic development is imbalance in Beijing. In terms of GRP, Central City is 1,056 billion CNY in 2010, which accounts for 74.59 % of total GRP in Beijing (Table 2.2).

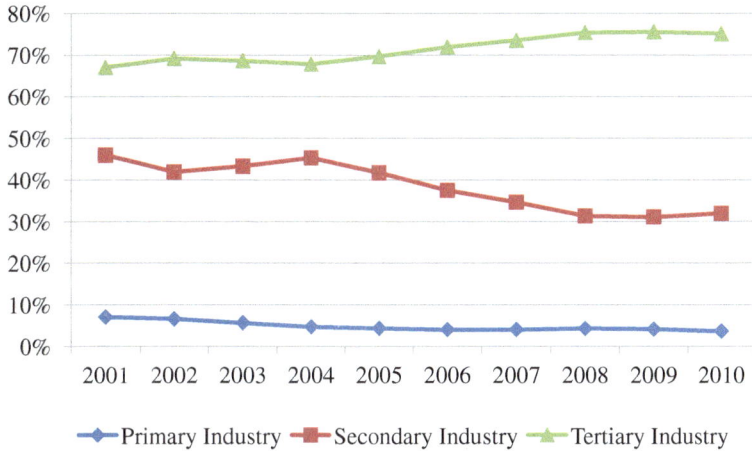

Fig. 2.3 GRP structure trend from 2001 to 2010 in Beijing

Table 2.2 GRP of every subregion in billions

Subregion	GRP	Percentage (%)
Central City	1,056	74.59
Fangshan	37	2.62
Tongzhou	34	2.44
Shunyi	87	6.13
Changping	40	2.83
Daxing	101	7.14
Mentougou	9	0.61
Huairou	15	1.05
Pinggu	12	0.83
Miyun	14	1.00
Yanqing	11	0.76
Total	1,415	100

2.5 Finance Revenue and Expenditure

Local financial revenue of the city totaled 381 billion CNY and saw an increase of 42 % over 2009 and an average of over 25 % increase for a consecutive of 10 years. Beijing's local expenditure was 406 billion CNY, which equaled to an increase of 6 % over the previous year. Over the past decade (2001–2010), Beijing fiscal revenue and expenditure are shown in Fig. 2.4.

In term of districts and counties financial revenue, Central City is 81 billion in 2010, which is also the highest amount of all subregions. Finance revenue and expenditure of every subregion is show in Table 2.3.

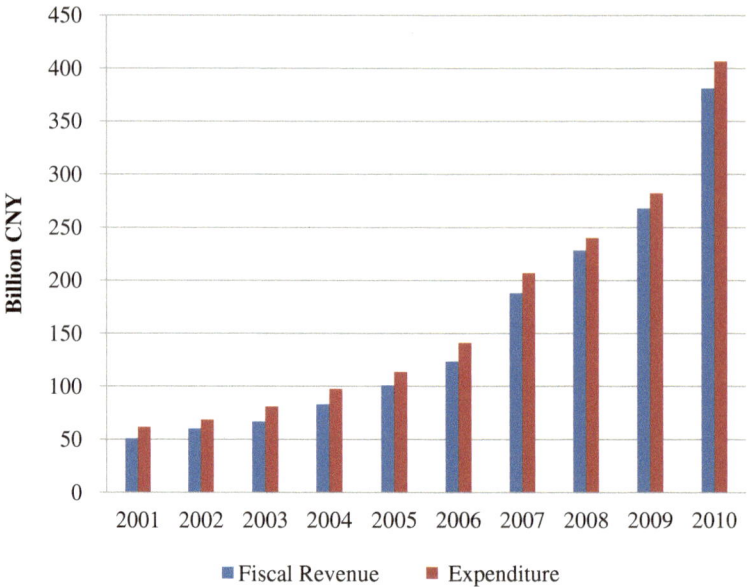

Fig. 2.4 Beijing fiscal revenue and expenditure trend from 2001 to 2010

Table 2.3 Finance revenue and expenditure of every subregion in 2010 in billions

Subregion	Finance revenue	Expenditure
Central City	81	103
Fangshan	17	22
Tongzhou	17	18
Shunyi	15	16
Changping	14	14
Daxing	26	30
Mentougou	3	6
	2	8
Huairou Pinggu	4	8
Miyun	2	8
Yanqing	1	6

2.6 Sewage Generation and Treatment

Due to its rapid economic and population growth, Beijing's municipal sewage generations are increasing each year. In 2010, the city produced more than 1.4 billion tons of sewage emissions (Beijing Environmental Protection Bureau 2011). Many sewage treatment plants that have been constructed by the Beijing municipal government have adopted advanced technologies, and the sewage treatment rate

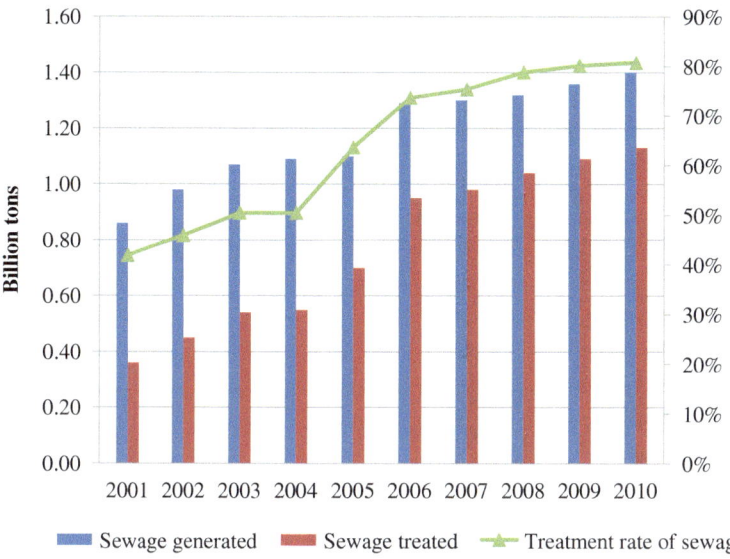

Fig. 2.5   Sewage generated and treated situation for Beijing

Table 2.4   Information of sewage generation and treatment of every subregion in 2010

Subregion	Sewage generation (million tons)	Sewage treatment (million tons)	Rate of sewage treatment (%)
Central City	810	779	96
Fangshan	52	36	69
Tongzhou	84	51	61
Shunyi	79	40	51
Changping	120	90	75
Daxing	95	68	72
Mentougou	40	14	35
Huairou	22	9	41
Pinggu	32	19	59
Miyun	46	16	35
Yanqing	45	11	24
Beijing	1,425	1,133	80

increased to 80 % in 2010 (Appendix 1; Beijing Water Authority 2011). Recent years, sewage generation and treatment conditions are shown in Fig. 2.5.

In term of subregion, the amount of sewage is 784 million tons in Central City, which account for 55 % of total sewage generate in Beijing. And the rate of sewage treatment is 96 %, which is much higher than other subregions. The detail information of sewage generation and treatment is shown in Table 2.4.

2.7 Sewage Sludge Generation and Treatment Capacity

As the byproduct of sewage treatment, the amount of sewage sludge also increases every year (Fig. 2.6). In 2010, the amount of sewage sludge is 1.13 million tons. However, the capacity of sewage treatment is only 0.48 million tons (Tables 2.5 and 2.6). In fact the rate of sewage treatment is less than 30 %. Therefore, there is amount of sewage untreated generated every year. If this sewage sludge cannot be treated, 50 % of the water pollutants removed by sewage treatment will return to the environment (Yang 2010). However, the need for sewage sludge treatment has not been addressed by the government.

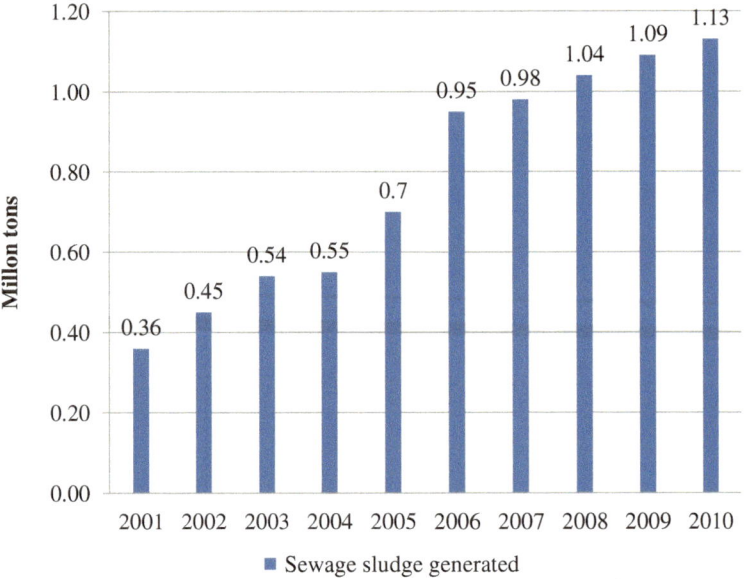

Fig. 2.6 Sewage sludge generated in Beijing

Table 2.5 The capacity of sewage treatment of Beijing in 2010 (thousand tons/year)

No	District	Name of Sludge Plant	Capacity
1	Daxing	Pang Ge Zhuang Composting Plant	110
2	Chang Ping	Chang Ping Composting Plant	29
3	Feng Tai	Fang Zhuang Limestone Drying Plant	11
4	Haidian	Qing He Sludge Heat Drying Plant	146
5	Changping	Beijing Cement Plant	184

Sources Tan et al. (2011), pp.105–109

Table 2.6 Sewage sludge generation and treatment capacity of every subregion in 2010

Subregion	Sewage sludge generation (thousand tons)	Capacity of sewage sludge treatment (thousand tons)
Central City	737	157
Fangshan	41	0
Tongzhou	51	0
Shunyi	40	0
Changping	104	213
Daxing	68	110
Mentougou	15	0
Huairou	21	0
Pinggu	29	0
Miyun	16	0
Yanqing	11	0
Beijing	1,133	480

2.8   Water Quality

In 2010, water shortages (Fig. 2.7) and water pollution in urban river downstream represent a serious situation that still has not been fundamentally reversed. The river water quality in Beijing is bad compared with other types of water class; 43.2 % of the river water is considered to be "inferior V" class, which is the worst level in China. 6.3 % of lake water is considered to be "inferior V" class. The water quality class proportion of rivers, lakes and reservoir is shown in Table 2.8. The surface water quality classification standard is shown in Table 2.7.

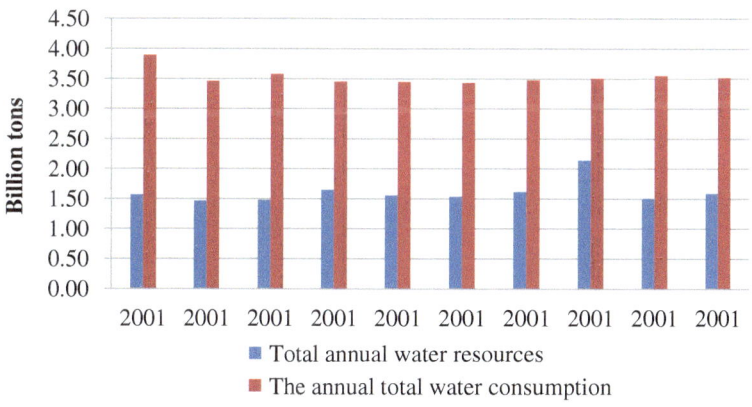

Fig. 2.7 The amount of water resource and water consumption trend in Beijing

Table 2.7 Surface water quality classification standards in China (mg/L)

	I	II	III	IV	V
COD	≤15	≤15	≤20	≤30	≤40
NH$_3$–N	≤0.15	≤0.5	≤1.0	≤1.5	≤2.0
T–P	≤0.02	≤0.1	≤0.2	≤0.3	≤0.4
T–N	≤0.2	≤0.5	≤0.1	≤1.5	≤2.0

Source Ministry of Environmental Protection of PRC's National Environmental Quality Standers for Surface Water (GB3838-2002), 2002, Beijing, China

Table 2.8 The water quality class proportion of rivers, lakes, and reservoir (%)

	II	III	IV	V	inferior V
River	49.1	6.4	0.2	1.1	43.2
Lake	31.0	45.2	14.2	3.3	6.3
Reservoir	85.8	3.7	10.5	–	–

Fig. 2.8 Load of COD in Beijing from 2001 to 2010

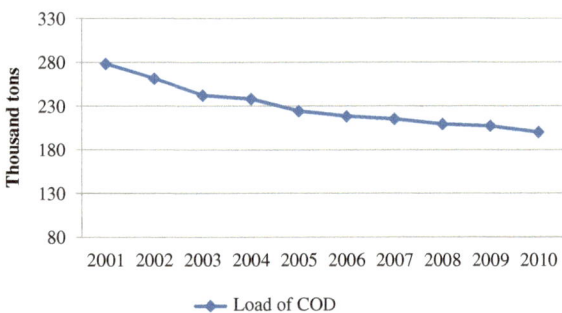

— Load of COD

Figure 2.8 Shows load of COD in Beijing from 2001 to 2010. It is declining since the government has put forward proposals to improve water quality. However, COD is still the most important water pollutant which is to be reduced to improve water quality. The government has planned that COD should be reduced 8.7 % in 2015 compared with in 2010.

2.9 Energy Consumption and GHG Emission

Figure 2.9 shows total energy consumption and energy consumption intensity from 2001 to 2020. The trend of energy consumption is increasing, but the energy consumption intensity is decreasing. According to Beijing government's plan, energy consumption intensity should be reduced 17 % in 2015 compared to in 2010 to achieve the GHG reduction target in 2015.

GHG emission by energy consumption is about 110 million tons in 2010. GHG emission intensity is about 78,052 tons/billion CNY. According to *The Twelfth Five-Year Plan of Economic and Social Development* (Beijing Municipal

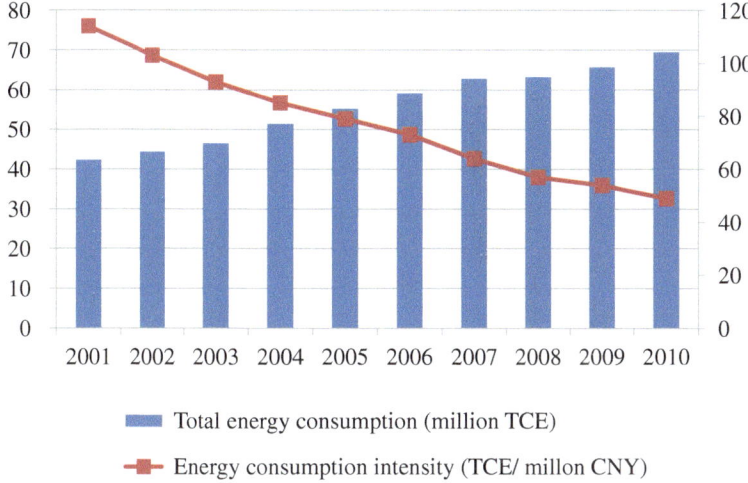

Fig. 2.9 Energy consumption and energy intensity from 2001 to 2020

Development and Reform Commission 2011), GHG emission intensity should be reduced 18 % in 2015 compared to in 2010. And in order to achieve the reduction target of GHG emission in 2020 of China, GHG emission intensity of Beijing must be reduced 36 % in 2020 compared to in 2010, while energy consumption intensity must be reduced 34 % in 2020 compared to in 2010.

2.10 Implications and Conclusions

Based on the current analysis of Beijing city, we can get the conclusion that, due to its rapid economic and population growth, water pollution, greenhouse gas emissions, energy demand increase is the key problems which is constraining economic and environmental sustainable development of Beijing.

Recently, the government has realized the importance of environmental protection. Accordingly, The Twelfth Five-Year Plan of Economic and Social Development (Beijing Municipal Development and Reform Commission 2011) requires that all sewage sludge be treated by 2015 and load of COD be reduced by 8.7 % in 2015 compared with 2010. In addition, in order to achieve the reduction target of GHG emission in 2020 of China, GHG emission intensity of Beijing must reduce 36 % in 2020 compared to in 2010, while energy consumption intensity reduced 34 % in 2020 compared to in 2010.

However, there is a big gap between the current situation and the future target. In order to realize these targets, Beijing government has adopted an integrated policy that are forestation for water conservation, reduction of working capital, and the introduction of advanced sewage sludge treatment technologies. However, how to evaluate the economic and environmental impact of the integrated policy is an urgent issue for Beijing government. To determine the optimal development plan for Beijing, it is beneficial to use a simulation method to evaluate the regional environmental and economic impacts.

References

Beijing Environmental Protection Bureau (2011) Beijing Environment Bulletin 2010 [EB/OL]: http://www.bjepb.gov.cn/portal0/tab181/, 3 June 2011 (in Chinese)

Beijing Municipal Development and Reform Commission (2011) The Twelfth Five-Year Plan for the National Economic and Social Development of Beijing [EB/OL]: http://www.bjpc.gov.cn/fzgh_1/guihua/12_5/Picture_12_F_Y_P/, 21 Jan 2011

Beijing Water Authority (2011) Beijing Water Resources Bulletin 2010 [EB/OL]: http://www.bjwater.gov.cn/tabid/207/Default.aspx, 8 March 2011 (in Chinese)

Tan GD, Li WZ, He CL (2011) Preliminary discussions on the sludge treatment and disposal technology in urban municipal sewage treatment plants in Beijing. South to North Water Divers Water Sci Technol 9(2):105–109 (in Chinese)

Yang XP (2010) Status and ideas of the Beijing Municipal sludge disposal. Water Wastewater Inf 8:17–18 (in Chinese)

Chapter 3
Comprehensive Evaluation Model of Socio-Economic and Environmental Policies

Abstract In this chapter, a comprehensive optimization simulation model was constructed which has four submodels that are: water pollutant balance model, energy model, GHG model, and socio-economic model. The objective function is the maximization of economic growth. Integrated policies by using advanced sewage and sewage sludge treatment technologies are dynamically simulated in this model. Based on this simulation results, optimal integrated policies, sewage, and sewage sludge treatment technologies can be selected. The environmental and economic impact of sewage sludge treatment can be forecasted.

Keywords Model · Simulation · Integrated policies · Advanced technologies

Accordingly, the Twelfth Five-Year Plan of Economic and Social Development (Beijing Municipal Development and Reform Commission 2011) requires that all sewage sludge should be treated by 2015 and load of COD (chemical oxygen demand) be reduced by 8.7 % in 2015 compared with 2010. Therefore, the government has adopted an integrated policy that includes forestation for water conservation, reduction of working capital, and the introduction of advanced sewage sludge treatment technologies.

A preliminary work has proved that the dynamic simulation method is reliable and the model is accurate in estimating the future economic development of the local area, and that COD should be selected as a limiting factor when advanced sewage sludge technology is adopted (Zhang 2013). However, advanced technology for sewage sludge treatment can not only remove the water pollutants but also can produce power and reduce greenhouse gas (GHG) emission. If the methane emitted in the anaerobic digestion process of sludge treatment is used to produce power, the reduction rate of GHG is 89 % (Li et al. 2011). Therefore, we should comprehensively consider the economic impact, water pollutants reduction, GHG reduction, and energy production for sewage sludge treatment.

To determine the optimal development plan for Beijing, it is beneficial to use a comprehensive simulation method to evaluate the regional environmental and economic impacts of adopting advanced technologies for the treatment of sewage sludge.

© The Author(s) 2016   29
G. Zhang, *Environmental and Social-economic Impacts of Sewage Sludge Treatment*, SpringerBriefs in Economics, DOI 10.1007/978-981-287-948-6_3

3.1 Concepts of Comprehensive Simulation Model

In this study, we draw on the research of Higano and Sawada (1997), Higano and Yoneta (1999), Hirose and Higano (2000), Mizunoya et al. (2007), and Yan (2010). Based on these previous studies, we constructed a comprehensive linear programming model. Moreover, this study demonstrates improvements over previous water pollutant models, energy models, and GHG models by introducing advanced sludge treatment technologies into the model. This comprehensive model consists of one objective function (Maximize GRP) and four submodels (a water pollutant balance model, an energy model, a GHG model and a socio-economic model) (Fig. 3.1). The economic model describes the relationship of economic activity, the emission of water pollutants, the consumption of energy, and the emission of GHG. The water pollutant model depicts changes in the level of water pollutants generated. The pollutants measured in this study are total nitrogen (T–N), total phosphorus (T–P), and chemical oxygen demand (COD). The energy model represents the relationship between energy demand (consumption by industry, new technology, and final consumption) and supply (from energy production enterprises and

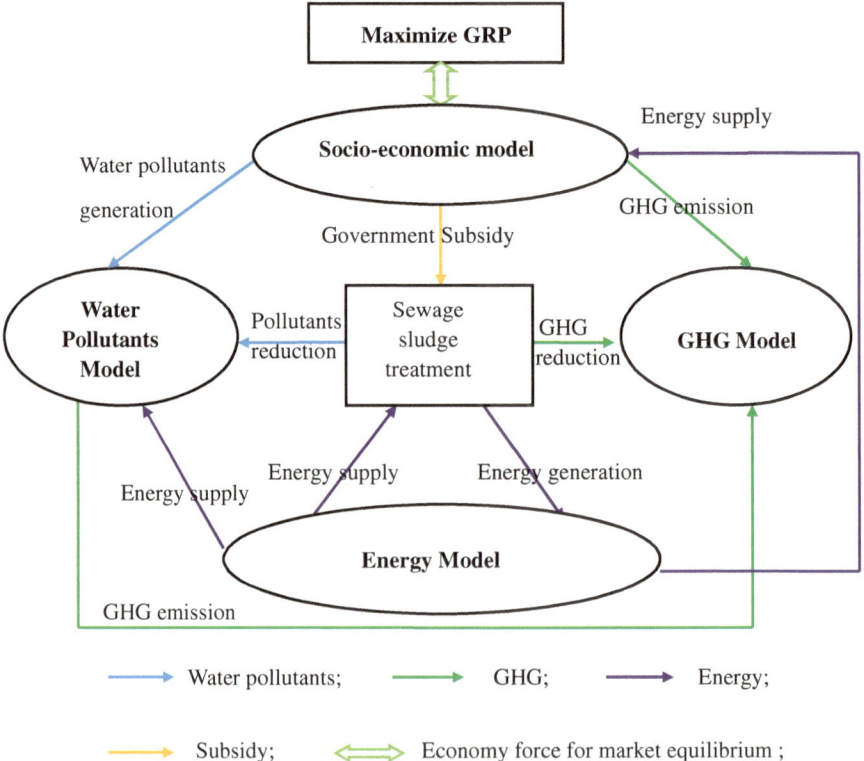

Fig. 3.1 The framework of the comprehensive simulation model

Table 3.1 Classification of water pollutants

Index	Water Pollutants
1	T–P
2	T–N
3	COD

dispatch). The energy measured is tons of standard coal (TCE). The GHG model set up the relationship between energy consumption and GHG emission. The GHGs measured in this study are carbon dioxide (CO_2), methane (CH_4), and nitrous oxide (N_2O). The simulation duration is from 2010 to 2020.

The evaluation indicators used in this work are economy, environment, and energy. These indicators are described by GRP, total load water pollutions, GHG emission intensity (GHG emission per billion GRP), and energy consumption intensity (energy consumption per billion GRP), respectively.

3.1.1   Classification of Water Pollutants

T–P, T–N, and COD are commonly used in the literature to describe water pollution (Kyou et al. 1998; Chae et al. 2007; Wang et al. 2008; Jing et al. 2009). Thus, the model in this study includes organic pollution parameters that are commonly used in the literature to describe water pollution. T–P, T–N, and COD coded as water pollutant 1, 2, and 3, respectively, in this analysis (Table 3.1).

3.1.2   Classification of Water Pollutant Sources

The pollutant sources are divided into three categories: nonpoint, household, and industry. Load of water pollutants via nonpoint source is decided by land area and the coefficient of water pollutant emissions for different land use. Land use is categorized into four types based on the "Beijing year book 2011" (Beijing Municipal Bureau of Statistics 2012), (Table 3.2), and industry is divided into fifteen categories based on "Beijing input-output extension Table 2010" (Beijing Municipal Bureau of Statistics 2011), (Table 3.3).

Table 3.2 Coding of land use types

No	Land use
1	Agricultural area
2	Forest
3	Urban area
4	Other land use

Table 3.3 Industry classification and coding

No	Industry
1	Agriculture
2	Forestry
3	Animal Husbandry
4	Fishery
5	Minerals Mining
6	Processing of Petroleum, Coking and Processing of Nuclear Fuel
7	Chemical Industry
8	Metallurgical Industry
9	Iron and Steel Industry
10	Manufacture of Communication Equipment, Computers and Other Electronic Equipment
11	Other Manufacturing
12	Production and Distribution of Electric Power and Heat
13	Construction
14	Transportation, Warehousing and Postal Service
15	Other services

3.1.3 Subregions of Beijing

To facilitate the collection of data and the implementation of policies, in this analysis Beijing is divided into 11 regions based on administrative divisions. The Central City subregion of Beijing, which contains six districts, is regarded as one region because some districts, such as Dongcheng and Xicheng, share sewage and sewage sludge treatment facilities with other districts. The other ten districts each represent a separate subregion (Table 3.4 and Fig. 3.2).

Table 3.4 Subdivision of the study area

No	Subregion	No	Subregion
1	Dongcheng	4	Shunyi
	Xicheng	5	Changping
	Haidian	6	Daxing
	Chaoyang	7	Mentougou
	Shijingshan	8	Huairou
	Fengtai	9	Pinggu
2	Fangshan	10	Miyun
3	Tongzhou	11	Yanqing

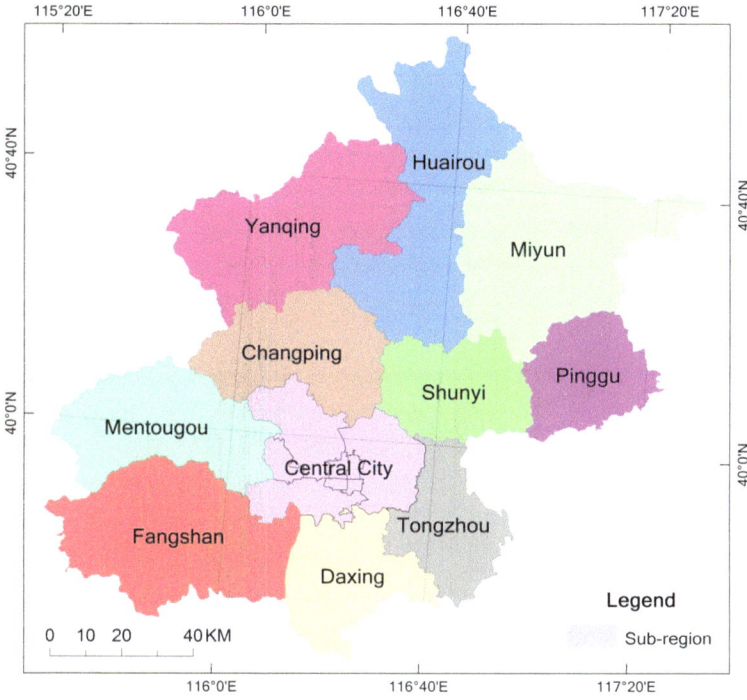

Fig. 3.2 The map of Beijing administrative division

Table 3.5 Classification of GHG

No	GHG	Potential index of greenhouse effect
1	CO_2	1
2	CH_4	25
3	N_2O	298

3.1.4   Classification of GHG

In this research, the GHG emission is measured taking into account CO_2, CH_4, and N_2O gases. CH_4 and N_2O have a much greater greenhouse effect potential than CO_2. Therefore, in order to appropriately evaluate emission of greenhouse gas, we have to consider the greenhouse effect potential of each gas (Table 3.5).

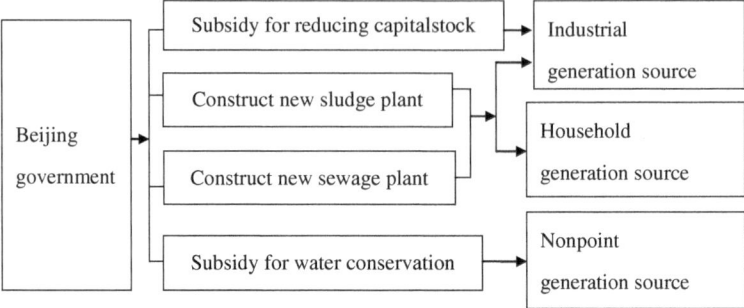

Fig. 3.3 Integrated policy

3.1.5 Integrated Policy for Water Pollutant Reduction

In this study, we propose an integrated policy to achieve a sustainable waste water management, in both the environmental and economic senses (Fig. 3.3). Currently, the policies used to reduce water pollutants are named "Forestation for water conservation" and "Reduction of working capital." "Forestation for water conservation" policy is that using government subsidies to encourage foresting on vacant land to enhance soil and water conservation. "Reduction of working capital" policy is that the exploitation of decrease of working capital for achieving water pollutant emission reduction by degrading the sectors with high emissions. In order to further reduce pollution of waterways, we propose the construction of new sewage and sewage sludge treatment plants to manage untreated sewage and sewage sludge in this study. Those are the policies of subsidy for reducing capital stock and subsidy for water conservation, which can decrease the pollution from industry and non-point source, respectively, and the policy of installation of new sewage and sewage sludge treatment plants, which can contribute to pollution reduction caused by industry and household.

The policy variables are represented by coefficients in the simulation model, and the environmental and economic impact of different combinations of policies are discussed based on the simulation results.

3.1.6 New Technology

We introduce four types of new sewage treatment technologies and four types of new sewage sludge treatment technologies. The new sewage treatment technologies include membrane bioreactor (MBR), dynamic membrane bioreactor (DMBR), ultrasonic membrane bioreactor (UMBR), and extractive membrane bioreactor (EMBR) (Table 3.6). The new sewage sludge treatment technologies include Anaerobic Digestion Fluidized bed drying I (A–D–F–I) is that it has been used in

Table 3.6 New sewage treatment technologies

	Investment (million CNY)	Operation cost (CNY/ton)	Capacity of water treatment (million tons/year)	Influent (mg/L)			Effluent (mg/L)		
				T–P	T–N	COD	T–P	T–N	COD
MBR	50	1.5	18	6	60	450	0.5	15	30
DMBR	165	2.0	37	6	65	450	1.0	10	15
UMBR	70	2.7	11	6	65	500	1.0	10	15
EMBR	5	1.8	1.5	6	80	600	1.0	8	40

Table 3.7 New sewage sludge treatment technologies

Technical Route	Investment (CNY/ton)	Operation Cost (CNY/ton)	Capacity (thousand tons)	Biogas Production (m³/ton)	Power Generation (kwh/ton)	GHG Emission Reduction (kg/ton)
A–D–F-I	526	714	50	138	–	274
A–D–F-II	556	833	50	218	–	425
F–C-I	500	1,250	50	–	277	276
F–C-II	667	1,000	50	–	266	259

Dalian, a city in Liaoning province; Anaerobic Digestion Fluidized bed drying II (A–D–F-II) that has been used in Shanghai City; Fluidized bed drying-Combustion I (F–C-I) that has been used in Jiaxing City in Zhejiang province; Fluidized bed drying-Combustion II (F–C-II) that has been used in Qingdao City in Shandong province (Table 3.7).

3.2 Objective Function

The maximization of the objective function, which gives the dynamic force to socio-economic system and simulates market equilibrium, which is described by the following structural equations:

$$Max \sum_t \frac{1}{(1+\rho)^{(t-1)}} GRP(t), \qquad (3.1)$$

$$GRP(t) = \sum_m v^m \cdot X^m(t), \qquad (3.2)$$

in which

ρ: social discount rate which is a measure used to help guide choices about the value of diverting funds to social projects (ex.), $\rho = 0.05$;

$GRP(t)$: Gross regional product (en.);

v: the row vector of the mth element that is the rate of added value in the mth industry (ex.);

$X^m(t)$: the column vector of the mth element is the total product of industry m in the target area at time t (en.).

3.3 Water Pollutants Model

3.3.1 The Framework of the Water Pollutants Model

We assumed that water pollutants for economic activity flow into the rivers in Beijing (Fig. 3.4). There are three sources of water pollutant discharge from economic activity: household, industry, and nonpoint sources. Herein, the pollutants contained in rainfall have been examined separately as one part of total water pollutant emissions in the City of Beijing regardless of its amount are very small (Hirose et al. 2000; Mizunoya et al. 2007; Yan 2010). Consequently, the pollutants emitted via rainfall were excluded from nonpoint sources. A portion of sewage flows into sewage plants through pipes, and sewage sludge is produced during the

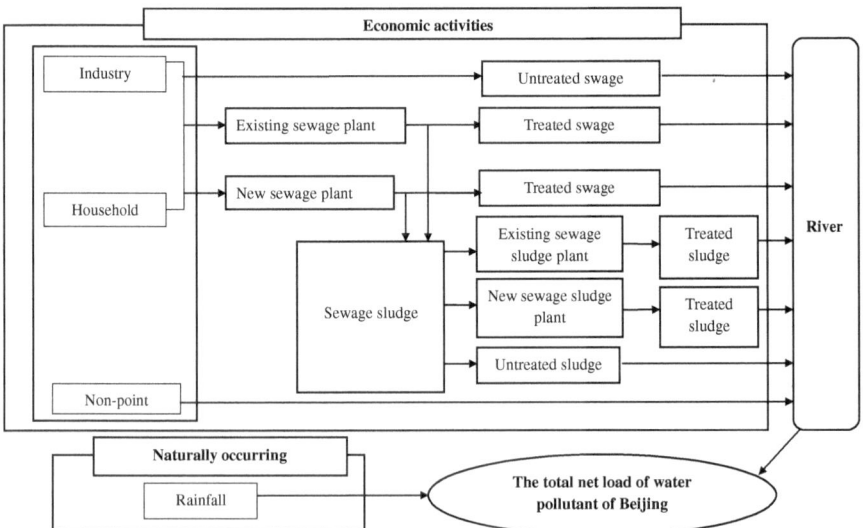

Fig. 3.4 The framework of water pollutant model

process of sewage treatment. Finally, the water pollutants contained in the treated and untreated sewage and sewage sludge flow into the rivers. Since it is difficult to collect the water pollutants added by nonpoint sources, we assume that these flow into rivers directly.

3.3.2 Specific Formulation of Water Pollutants Model

(1) Total water pollutant load for Beijing

The total water pollutant load in Beijing is simulated as the sum of water pollutant of every subregion. Total water pollutant load in Beijing is described in this equation

$$TQ^p(t) = \sum_j WP_j^p(t), \ (p = 1 : T-P; p = 2; T-N; p = 3 : COD) \qquad (3.3)$$

in which

$TQ^p(t)$: total load of water pollutant p for Beijing at time t (en.);
$WP_j^p(t)$: load of water pollutant p in region j at time t (en.).

(2) The constraints for the total water pollutant load for Beijing

In order to achieve Beijing government's plan, we set a boundary of water pollutant load which is calculated based on the real data in 2010. In this simulation, the total load of water pollutant should equal or less than the boundary. We express in the following equation:

$$TQ^p(t) \leq TQC^p(t), \qquad (3.4)$$

in which

$TQC^p(t)$: the boundary of water pollutant p at time t (en.).

(3) Water pollutant load of subregion

Load of water pollutant in subregion is simulated as the sum of the flows from rivers and rainfall. We expressed it in the following equation:

$$WP_j^p(t) = QR(t) + RQ_j^p(t), \qquad (3.5)$$

in which

$QR(t)$: load of water pollutant p in rivers at time t (en.);
$RQ_j^p(t)$: load of water pollutant p from rainfall at time t (en.).

(4) pollutant flow through rivers

Water pollutant contributed by economic activities flow into revivers. And then, part of water pollutant removed because rivers have the capacity of self-purification. Load of water pollutant in rivers is described as

$$QR_j^p(t) = (1 - \upsilon) \cdot SECQ_j^p(t), \tag{3.6}$$

in which

$QR(t)$: load of water pollutant p in rivers at time t (en.);
υ: river self-purification rate (ex.);
$SECQ_j^p(t)$: water pollutant p contributed by economic activities in the region j at time t (en.).

(5) Water pollutant contributed by economic activities

Water pollutant sources of economic activities are divided into three categories, which are households, industry and nonpoint. Some of the water pollutants emitted by economic activities is removed by the sewage and sewage sludge treatment. Water pollutant contributed by economic activities is described as

$$SECQ_j^p(t) = HQ_j^p(t) + UIQ_j^p(t) + NQ_j^p(t) - SEQ_j^p(t) - SLQ_j^p(t), \tag{3.7}$$

in which

$HQ_j^p(t)$: water pollutant p emitted by households in region j at time t (en.);
$UIQ_j^p(t)$: water pollutant p emitted by industry in region j at time t (en.);
$NQ_j^p(t)$: water pollutant p emitted by nonpoint sources in region j at time t (en.);
$SEQ_j^p(t)$: water pollutant p reduced by sewage plants in region j at time t (en.);
$SLQ_j^p(t)$: water pollutant p reduced by sewage sludge plants in region j at time t (en.).

(6) Load of water pollutants from nonpoint sources

Water pollutant through nonpoint sources is decided by the land area and the emotion coefficient of water pollutant of different land use. Water pollutants emitted by nonpoint sources is set as

$$NQ_j^p(t) = \sum_g EL^g \cdot L_j^g(t), \tag{3.8}$$

in which

EL^g: coefficient of water pollutant p emitted by land use g (ex.);
$L_j^g(t)$: area of land use g in region j at time t (en.).

(7) Water pollutants emitted by households

Water pollutant emitted by household is decided by the number of household and the emission coefficient of water pollutant per household with the following equations:

$$HQ_j^p(t) = Z_j(t) \cdot EH^p, \tag{3.9}$$

$$Z_j(t+1) = Z_j(t) \cdot (1 + \mu), \tag{3.10}$$

in which

$Z_j(t)$: number of households in region j at time t (en.);
EH^p: emission coefficient of water pollutant p per household (ex.);
μ: household growth rate (ex.).

(8) Water pollutants emitted by industry

Level of water pollutants for industry is dependent upon production. We describe relationship of production and emission of water pollutant via coefficient of water pollutants emissions of industry. This relationship is set as

$$UIQ_j^p(t) = \sum_m x_j^m(t) \cdot EUI^m, \tag{3.11}$$

in which

$x_j^m(t)$: production of industry m in region j at time t (en.);
EUI^m: emission coefficient of water pollutant p of industry m (ex.).

(9) Water pollutant reduced by sewage plants

Water pollutant reduced by sewage plants has two portions. One portion is reduced by existing sewage plants. Nine types of technology have been use for these sewage plants (Appendix A). Another portion is reduced by new sewage plants, which will use two types of advanced technologies. Water pollutant reduced by sewage plants is described as

$$SEQ_j^p(t) = SEQ_j^a(t) + SEQ_j^b(t), \tag{3.12}$$

$SEQ_j^a(t)$: load of water pollutant p reduced by the existing sewage plants using original technology a in region j at time t (en.);
$SEQ_j^b(t)$: load of water pollutant p reduced by the new sewage plants, which use advanced technology b in region j at time t (en.).

Water pollutant reduced by existing sewage plants is described as

$$SEQ_j^a(t) = \sum_a QSE_j^a(t) \cdot \alpha^a, \tag{3.13}$$

in which

$QSE_j^a(t)$: the amount of sewage treated by existing plants, which use original technology a, in region j at time t (ex.);
α^a: coefficient of reduction of pollutant p by original sewage technology a (ex.).

Water pollutant reduced by new sewage plants is described as

$$SEQ_j^b(t) = \sum_b QSE_j^b(t) \cdot \alpha^b, \tag{3.14}$$

in which

$QSE_j^b(t)$: the amount of sewage treated by new sewage plants, which use advanced technology b, in region j at time t (en.);
α^b: coefficient of reduction of pollutant p by advanced sewage technology b (ex.).

(10)   Water pollutant reduced by sewage sludge plants
Water pollutant reduced by sewage sludge plants has two portions. One portion is reduced by existing sewage sludge plants. Five types of technologies have been used for these sewage plants. Another portion is reduced by new sewage sludge plants, which will use two types of advanced technologies. Water pollutant reduced by sewage sludge plants is described as

$$SLQ_j^p(t) = SLQ_j^c(t) + SLQ_j^d(t), \tag{3.15}$$

in which

$SLQ_j^c(t)$: load of water pollutant p reduced by the existing sewage sludge plants, which use original technology c, in region j at time t (en.);
$SLQ_j^d(t)$: load of water pollutant p reduced by the new sewage sludge plants, which use advanced technology d, in region j at time t (en.).

Water pollutant reduced by existing sewage sludge plants is described as

$$SLQ_j^c(t) = \sum_c ES^c \cdot QSL_j^c(t), \tag{3.16}$$

in which

ES^c: coefficient of reduction of pollutant p by original sludge technology c (ex.);

$QSL_j^c(t)$: amount of sewage sludge treated by existing plants, which use original technology c, in region j at time t (ex.).

Water pollutant reduced by new sewage sludge plants is described as

$$SLQ_j^d(t) = \sum_d ES^d \cdot QSL_j^d(t), \qquad (3.17)$$

ES^d: coefficient of reduction of pollutant p by new sludge technology d (ex.); $QSL_j^d(t)$: amount of sewage sludge treated by new plants, which use advanced technology d in region j at time t (en.).

(11)   Load of water pollutants from rainfall
The load of water pollutant of the rainfall is given as the following equation:

$$RQ_j^p(t) = ER^p(t) \cdot L_j(t), \qquad (3.18)$$

in which

ER^p: emission coefficient of rainfall for pollutant p (ex.), $ER^1 = 47$ kg/ km^2-year, $ER^2 = 1{,}124$ kg/ km^2-year, and $ER^3 = 2{,}091$ kg/ km^2-year (Hirose et al. 2000; Yan 2010);
$L_j(t)$: total area of region j at time t (ex.).

3.4   Energy Model

3.4.1   Framework of Energy Model

The energy model draws the relationship between energy demand and supply (Fig. 3.5). Energy is demanded by industry production, sewage treatment, sewage sludge treatment and final consumption. Energy is supplied by energy production enterprises, power produced by sewage sludge treatment and dispatch. In order to realize socio-economic development, energy supply must be equal or more than energy demand.

3.4.2   Specific Formulation of Energy Model

(1)   The total energy demand
Energy demand consists of industry, sewage treatment, sewage sludge treatment, and final consumption. The total energy demand is described as

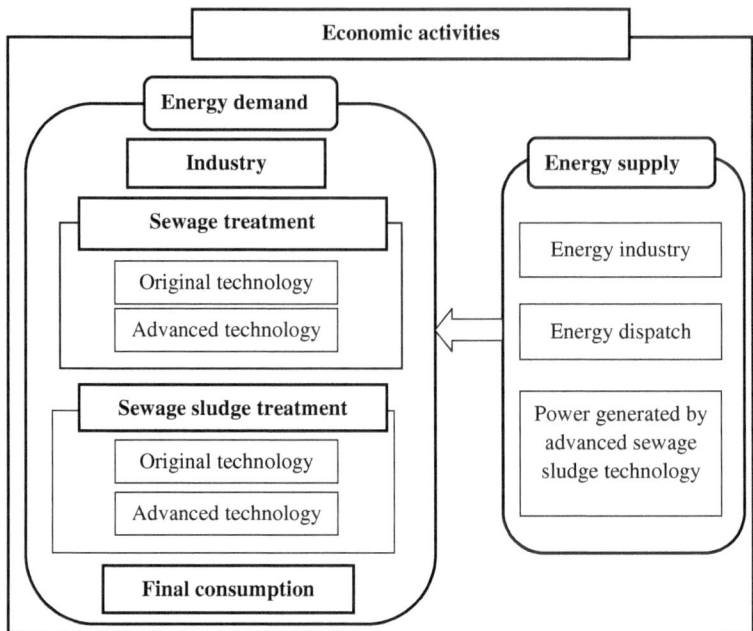

Fig. 3.5 The framework of energy model

$$TED(t) = IED(t) + SWED(t) + SLED(t) + FCED(t), \tag{3.19}$$

in which

$TED(t)$: the total amount of energy demand at time t (en.);
$IED(t)$: the amount of energy demand of industry at time t (en.);
$SWED(t)$:the amount of energy demand of sewage treatment at time t (en.);
$SLED(t)$: the amount of energy demand of sewage sludge treatment at time t (en.);
$FCED(t)$: the amount of energy demand of final consumption at time t (en.).

(2) The amount of energy demand of industry
Energy demand of industry is decided by the product of industry and coefficient of energy demand. The equation is set as

$$IED(t) = \sum \sum x_j^m(t) \cdot ED^m, \tag{3.20}$$

in which

$x_j^m(t)$: production of industry m in region j at time t (en.);
ED^m: coefficient of energy demand of industry m (ex).

(3) The amount of energy demand of sewage treatment

The amount of energy demand of sewage treatment is divided into two portions. One portion is energy demand by existing sewage plants; another portion is energy demand by new sewage plants.

The amount of energy demand of sewage treatment is as the following equations:

$$SWED(t) = SEWD^a(t) + SWED^b(t), \tag{3.21}$$

$$SWED^a(t) = \sum_j \sum_a QSE_j^a(t) \cdot EDC^a, \tag{3.22}$$

$$SWED^b(t) = \sum_j \sum_b QSE_j^b(t) \cdot EDC^b, \tag{3.23}$$

in which

$SWED^a(t)$:total amount of energy demand by existing sewage plants which uses technology a at time t (en);
$SWED^b(t)$: total amount of energy demand by new sewage plants which uses technology b at time t (en);
EDC^a: coefficient of energy demand of technology a (ex);
EDC^b: coefficient of energy demand of technology b (ex).

(4) The amount of energy demand of sewage sludge treatment

The amount of energy demand of sewage sludge treatment is divided into two portions. One portion is energy demand by existing sewage sludge plants; another portion is energy demand by new sewage sludge plants.

The amount of energy demand of sewage sludge treatment is as the following equations:

$$SLED(t) = SLED^c(t) + SLED^d(t), \tag{3.24}$$

$$SLED^c(t) = \sum_j \sum_c QSL_j^c \cdot EDC^c(t), \tag{3.25}$$

$$SLED^d(t) = \sum_j \sum_d QSL_j^d(t)EDC^d, \tag{3.26}$$

in which

$SLED^c(t)$: the total amount of energy demand by original sewage sludge treatment technology c at time t (en.);
$SLED^d(t)$: the total amount of energy demand by new sewage sludge treatment technology d at time t (en.);

EDC^c: the coefficient of energy demand of original sewage sludge technology c (ex.);

EDC^d: the coefficient of energy demand of new sewage sludge technology d (ex.).

(5) The amount of energy demand of final consumption

The amount of energy demand of final consumption is described as follow:

$$FCED(t) = C(t) \cdot FCC, \tag{3.27}$$

in which

$C(t)$: final consumption at time t (en.);

FCC: the coefficient of energy demand of final demand section (ex.).

(6) The total energy supply

Energy supply consists of energy industry, energy generation of sewage sludge treatment and energy dispatch. The total energy supply is described as follow:

$$TES(t) = EIS(t) + ESL(t) + DSP(t), \tag{3.28}$$

in which

$TES(t)$: the total amount of energy supply at time t (ex.);

$EIS(t)$: energy supply of energy industry at time t (ex.);

$ESL(t)$: energy generated by sewage sludge treatment (en.);

$DSP(t)$: the total amount of energy dispatch at time t (ex.).

Energy generation of sewage sludge treatment is set as

$$ESL(t) = \sum_j \sum_d QSL_j^d(t) \cdot EGC^d, \tag{3.29}$$

in which

EGC^d: the energy generation coefficient of new sewage sludge treatment technology d (ex.).

(7) Energy balance

In order to realize socio-economic development, energy supply must equal or more than energy demand. The equation is as follow:

$$TED(t) \leq TES(t), \tag{3.30}$$

in which

$TED(t)$: the total amount of energy demand at time t (en.);

$TES(t)$: the total amount of energy supply at time t (ex.).

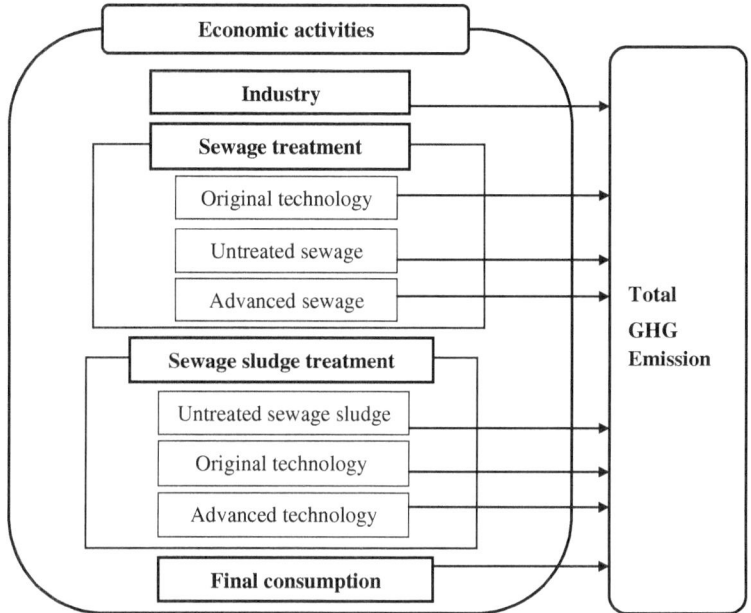

Fig. 3.6 The framework of GHG model

3.5   GHG Model

3.5.1   Framework of GHG Model

The GHG model gives a description of the relationship between energy consumption and GHG emission. GHG emission is determined by energy consumption. The sources of GHG emission are industry production, sewage treatment and final consumption (Fig. 3.6).

3.5.2   Specific Formulation of Energy Model

(1) Total CHG emission

In this research we select CO_2, CH_4 and N_2O as the greenhouse gas. In sum, the GHG emission sources consist of industry, sewage treatment, sewage sludge treatment and final consumption. Total CHG emission is described as the following equations:

$$TGHG(t) = TG^1(t) + TG^2(t) \cdot GWP^2 + TG^3(t) \cdot GWP^3, \qquad (3.31)$$

$$TG^a(t) = TGI(t) + TGSW(t) + TGSL(t) + TGFC(t), \\ (a = 1 : \ CO_2; a = 2 : \ CH_4; a = 3 : \ N_2O;) \qquad (3.32)$$

in which

$TGHG(t)$: total amount of GHG emission of Beijing at time t (en.);
$TG^a(t)$: the amount of GHG (a) emission at time t (en.);
$TGI(t)$: the amount of GHG emission of industry at time t (en.);
$TGSW(t)$:the amount of GHG emission of sewage at time t (en.);
$TGSL(t)$: the amount of GHG emission of sewage sludge at time t (en.);
$TGFC(t)$: the amount of GHG emission of final consumption (en.);
GWP: potential index of greenhouse effect.

(2) GHG emitted by industry
The amount of GHG emission of industry is decided by the product of industry and coefficient of GHG emission. The equation is set as

$$TGI(t) = \sum_j \sum_m x_j^m(t) \cdot EC^m, \qquad (3.33)$$

in which

$x_j^m(t)$: production of industry m in region j at time t (en.);
EC^m: coefficient of greenhouse gas of industry m (ex.).

(3) GHG emitted by sewage
Amount of GHG emitted by sewage is divided into three portions. Fist portion is emitted by existing sewage plants; second portion is emitted by new sewage plants; third portion is emitted by untreated sewage. The amount of GHG emitted by sewage is as the following equations:

$$TGSW(t) = TGSW^a(t) + TGSW^b(t) + TGSW^u(t), \qquad (3.34)$$

$$TGSW^a(t) = \sum_j \sum_a QSE_j^a(t) \cdot EC^a, \qquad (3.35)$$

$$TGSW^b(t) = \sum_j \sum_b QSE_j^b(t) \cdot EC^b, \qquad (3.36)$$

$$TGSW^U(t) = \sum_j QSET_j^U(t) \cdot EC^U, \qquad (3.37)$$

$TGSW^a(t)$: the amount of GHG emitted by original sewage technology a at time t (en.);

$TGSW^b(t)$: the amount of GHG emitted by new sewage technology b at time t (en.);

$TGSW^U(t)$: the amount of GHG emitted by untreated sewage at time t (en.);

EC^a: coefficient of greenhouse gas of original sewage technology a (ex.);

EC^b: coefficient of greenhouse gas of new sewage technology b (ex.);

EC^U: coefficient of greenhouse gas of untreated sewage (ex.).

(4) GHG emitted by sewage sludge

Amount of GHG emitted by sewage sludge is divided into three portions. Fist portion is emitted by existing sewage sludge plants; second portion is emitted by new sewage sludge plants; third portion is emitted by untreated sewage sludge. The amount of GHG emitted by sewage sludge is as the following equations:

$$TGSL(t) = TGSL^c(t) + TGSL^d(t) + TGSL^u(t), \qquad (3.38)$$

$$TGSL^c(t) = \sum_j \sum_c QSL_j^c(t) \cdot EC^c, \qquad (3.39)$$

$$TGSL^d(t) = \sum_j \sum_d QSL_j^d(t) \cdot EC^d, \qquad (3.40)$$

$$TGSL^U(t) = \sum_j QSL_j^U(t) \cdot EC^U, \qquad (3.41)$$

$TGSL^c(t)$: the amount of GHG emitted by original sewage sludge technology c at time t (en.);

$TGSL^d(t)$: the amount of GHG emitted by new sewage sludge technology d at time t (en.);

$TGSL^u(t)$: the amount of GHG emitted by untreated sewage sludge at time t (en.);

EC^c: coefficient of greenhouse gas of original sewage sludge technology c (ex.);

EC^d: coefficient of greenhouse gas of new sewage sludge technology d (ex.);

EC^U: coefficient of greenhouse gas of untreated sewage sludge (ex.);

(5) GHG emitted by final consumption

The amount of GHG emission of final consumption is described as follow:

$$TGFC(t) = C(t) \cdot EFC, \qquad (3.42)$$

in which

$C(t)$: final consumption at time t (en.);

EFC: coefficient of greenhouse gas emitted by final consumption (ex.).

3.6 Economic Model

3.6.1 Policy Treatment for Nonpoint Sources

The Beijing government converted land that into forest land (originally this land was assigned a different purpose), providing subsidies to improve the water quality. Total land area is expressed in this equation:

$$\bar{L}_j(t) = \sum_g L_j^g(t),$$
(3.43)

in which

$\bar{L}_j(t)$: total land area in region j at time t (ex.);
$L_j^g(t)$: land area comprised of land use g in region j at time t (en.).

The increasing area of forest land is deceived by government's subsidy. The changing area of forest land is described as follow equations:

$$L_j^g(t+1) = L_j^g(t) + \Delta L_j^g(t), \ (g = 2)$$
(3.44)

$$\Delta L_j^g(t) = L_j^{42}(t),$$
(3.45)

$$L_j^{42} \geq \lambda^4 \cdot S_j^{42}(t),$$
(3.46)

in which

$L_j^g(t+1)$: area of forest in region j at time $t+1$ (en.);
$\Delta L_j^g(t)$: increased area of forest that was converted from other land uses in region j at time t (en.);
$L_j^{42}(t)$: conversion from other land uses ($g = 4$) to forest ($g = 2$) in region j at time t (en.);
λ^4: reciprocal of the subsidy for one unit of conversion to forest (ex.);
$S_j^{42}(t)$: subsidy for region j given by the Beijing government for conversion of other land use into forest (en.).

3.6.2 Policy Treatment for Production Generation Sources

This production function is derived from Harrod-Domar model through the relationship between capital accumulation and production. The production of industry m is restricted by working capital and subsidy for reducing working capital (Yan 2010). Capital accumulation is dependent upon the investment and depreciation of capital.

The relationship between capital accumulation and production is set as

$$x_j^m(t) \le \alpha^m \left\{ k_j^m(t) - s_j^m(t) \right\}, \quad (m = 1, 2, \ldots, 15) \tag{3.47}$$

in which

$x_j^m(t)$: production of industry m in region j at time t (en.);
$k_j^m(t)$: working capital available for industry m in region j at time t (en.);
$s_j^m(t)$: subsidy given by the Beijing government for industry m at time t (en.);
α^m: ratio of capital to output in industry m (ex.).

The relationship between capital accumulation and investment is expressed as

$$k_j^m(t+1) = k_j^m(t) + I_j^m(t+1) - f^m \cdot k_j^m(t), \tag{3.48}$$

in which

$I_j^m(t+1)$: investment in industry m in region j at time $t+1$ (en.);
f^m: rate of capital depreciation of industry m (ex.).

3.6.3 Policy Treatment for New Sewage Plant Construction

The total amount of sewage is determined by the amount of industrial and household emissions. The total amount of sewage generation is expressed as

$$TQSE_j(t) = \sum_m x_j^m(t) \cdot \eta^m + Z_j(t) \cdot \eta^z, \tag{3.49}$$

in which

$TQSE_j(t)$: amount of sewage generated in region j at time t (en.);
$x_j^m(t)$: production of industry m in region j at time t (en.);
η^m: sewage emission coefficient for industry m (ex.);
$Z_j(t)$: number of households in region j at time t (en.);
η^z: household sewage emission coefficient (ex.).

The total amount of sewage treatment depends upon two quantities. The first amount is the amount of sewage treated by existing sewage plants, which uses original technologies, and the second is the amount of sewage treated by new sewage plants, which uses advanced technologies.

$$TQSET_j(t) = \sum_a QSE_j^a(t) + \sum_b QSE_j^b(t), \qquad (3.50)$$

in which

$TQSET_j(t)$: total amount of sewage treatment in region j at time t (en.);
$QSE_j^a(t)$: the amount of sewage treated by existing sewage plants, which use original technology a, in region j at time t (ex.);
$QSE_j^b(t)$: the amount of sewage treated by new sewage plants, which use advanced technology b, in region j at time t (en.).

The amount of sewage treatment must be equal or less than that of sewage generation which described as follow:

$$TQSET_j(t) \leq TQSE_j(t), \qquad (3.51)$$

in which

$TQSET_j(t)$: the amount of sewage treatment in region j at time t (en.).

The increase in the amount of sewage treatment depends upon the investment in new plant construction. The investment and maintenance cost of new plants is covered by subsidies from the Beijing government.

The increase in the amount of sewage treatment is expressed as the following equations:

$$TQSET_j(t+1) = TQSET_j(t) + \Delta TQSET_j(t), \qquad (3.52)$$

$$\Delta TQSET_j(t) = \sum_b \Delta QSE_j^b(t), \qquad (3.53)$$

$$QSE_j^b(t+1) = QSE_j^b(t) + \Delta QSE_j^b(t), \qquad (3.54)$$

$$\Delta QSE_j^b(t) \leq \Phi \cdot I_j^b(t), \qquad (3.55)$$

in which

$TQSET_j(t+1)$: total amount of sewage treatment in region j at time $t+1$ (en.);
$\Delta TQSET_j(t)$: increase in the quantity of sewage treatment in region j at time t (en.);
$\Delta QSE_j^b(t)$: quantity of sewage treatment increased by new sewage plants, which use advanced technology b, in region j at time t (en.);
$QSE_j^b(t+1)$: the amount of sewage treated by new sewage plants, which use advanced technology b, in region j at time $t+1$ (en.);
$I_j^b(t)$: investment in new sewage plants construction, which use advanced technology b, in region j at time t (en.);
Φ: sewage treatment coefficient per unit of investment (ex.).

The maintenance costs of new sewage plants depend upon the amount of sewage treatment by the new plants. The maintenance cost of new sewage plants is described as

$$MC_j^b(t) = \zeta_j^b \cdot QSE_j^b(t),\tag{3.56}$$

in which

$MC_j^b(t)$: maintenance costs of new sewage plants, which use technology b, in region j at time t (en.);
ζ_j^b: maintenance costs per ton of sewage for new sewage plants, which use technology b, in region j (ex.).

The investment and maintenance cost new sewage plants must be equal or less than the government's subsidy.

$$I_j^b(t) + MC_j^b(t) \leq S_j^b(t),\tag{3.57}$$

in which

$S_j^b(t)$: subsidies for new sewage plants construction, which use advanced technology b, in region j by Beijing government at time t (en.).

3.6.4 Policy Treatment for Sewage Sludge Construction

The total amount of sewage sludge is determined by the amount of sewage treatment and can be divided into two portions. One portion is the amount of sewage sludge treated by existing sewage sludge plants, which uses original technologies, and the other portion is the amount of sewage sludge treated by new sewage sludge plants, which uses advanced technologies.

Total amount of sewage sludge generation must meet the following equations:

$$TQSL_j(t) \leq TQSET_j(t) \cdot ESE,\tag{3.58}$$

$$TQST_j(t) \leq TQSL_j(t),\tag{3.59}$$

in which

$TQSL_j(t)$: total amount of sewage sludge generated d in region j at time t (en.);
$TQSET_j(t)$: total amount of sewage treatment in region j at time t (en.);
ESE: coefficient of sewage sludge generated by sewage treatment (ex.);
$TQSLT_j(t)$: the amount of sewage treatment in region j at time t (ex.).

The amount of sewage treatment is set as

$$TQSET_j(t) = \sum_c QSL_j^c(t) + \sum_d QSL_j^d(t),$$ (3.60)

in which

$TQSET_j(t)$: total amount of sewage sludge treatment in region j at time t (en.);
$QSL_j^c(t)$: amount of sewage sludge treated by existing sewage sludge plants, which use original technology c, in region j at time t (ex.);
$QSL_j^d(t)$: amount of sewage sludge treated by new sewage sludge plants, which use advanced technology d, in region j at time t (en.).

The increase in the amount of sewage sludge treatment is determined by investment in new sewage sludge plant construction. The increase in the amount of sewage sludge treatment is expressed by the following equations:

$$TQSET_j(t+1) = TQSET_j(t) + \Delta TQSET_j(t),$$ (3.61)

$$\Delta TQSLT_j(t) = \sum_d \Delta QSL_j^d(t),$$ (3.62)

$$\Delta QSL_j^d(t) \le \lambda_j^d \cdot I_j^d(t),$$ (3.63)

in which

$TQSLT_j(t+1)$: total amount of sewage sludge treatment in region j at time $t + 1$ (en.);
$\Delta TQSLT_j(t)$: increase in the amount of sewage sludge treatment in region j at time t (ex.);
$\Delta QSL_j^d(t)$: increase in the amount of sewage sludge treatment by new sewage sludge plants, which use advanced technology d, in region j at time t (en.);
λ_j^d: sewage sludge treatment coefficient per unit of investment for advanced technology d in region j (ex.);
$I_j^d(t)$: investment in new sewage sludge plants construction, which use advanced technology d, in region j at time t (en.).

The maintenance cost of new sewage sludge plants depends upon the amount of sewage sludge treatment by these new plants. Maintenance cost of new sewage sludge plants is set as

$$MC_j^d(t) = \zeta_j^d \cdot QSL_j^d(t),$$ (3.64)

in which

$MC_j^d(t)$: maintenance costs of new sewage sludge plants, which use advanced technology d, in region j at time t (en.);
ζ_j^d: maintenance costs per ton of sewage sludge treatment for new sewage sludge plants, which use advanced technology d, in region j (ex.).

The investment and maintenance cost of new sewage sludge plants must be equal or less than the government's subsidy.

$$I_j^d(t) + MC_j^d(t) \leq S_j^d(t), \tag{3.65}$$

in which

$S_j^d(t)$: subsidies for new sewage sludge plants construction, which use advanced technology d, in region j, from the Beijing government at time t (en.).

The subsidy of the integrated policy must meet the following equation:

$$S(t) \geq \sum_j \sum_m S_m^j + \sum_j S^{42}(t) + \sum_j \sum_b S_b^j + \sum_j \sum_d S_d^j, \tag{3.66}$$

in which

$S(t)$: total subsidy for water pollutant reduction from the Beijing government at time t (ex.).

3.6.5 Market Balance

The total production of each industry is determined by balances between supply and demand (Mizunoya et al. 2007). The production is dependent on the Leontief input–output coefficient matrix, consumption, investment, and net export (Yan 2010). We added variables that were related to the investment in advanced technologies to describe the impact of new plant construction on production. The equations are as follow:

$$X^m(t) \geq A \cdot X^m(t) + C(t) + i^m(t) + \beta_m^b I^b(t) + \beta_m^b I^b(t) + e(t), \tag{3.67}$$

$$I^b(t) = \sum_j \sum_b I_j^b(t), \tag{3.68}$$

$$I^b(t) = \sum_j \sum_d I_j^d(t), \tag{3.69}$$

$$X(t) = \sum_j x_j^m(t), \tag{3.70}$$

in which

$X(t)$: the column vector of the mth element is the total product of industry m in the target area at time t (en.);
A: input-output coefficient matrix (ex.);
$C(t)$: total consumption at time t (en.);
$i^m(t)$: total investment in industry m at time t (ex.);
β_m^b: the column vector of the mth coefficient is the production in industry m induced by new sewage plant construction(ex.);
$I^b(t)$: total investment in new sewage plant construction at time t (ex.);
β_m^d: the column vector of the mth coefficient is the production induced in industry m by new sewage sludge plant construction(ex.);
$I^d(t)$: total investment in new sewage sludge plant construction at time t (ex.);
$e(t)$: column vector of net export at time t (en.).

3.7   Case Setting

We defined water pollutant reduction rate as percentage of decrease in water pollutant in 2020 compared to 2010. Scenario 1 simulates the current situation, in
which advanced technology is not adopted. Scenario 2 simulates the effects of
implementation of integrated policies, which addresses both economic and environmental aspects, by introduction of advanced sewage treatment technologies.
Scenario 3 and 4 simulates effects of integrated policies in conjunction with
introduction of both advanced sewage and sewage sludge treatment technologies
(Table 3.8).

Table 3.8 Scenario composition

Scenarios	Water pollutant reduction rate (%)	Subsidy for water conservation	Reduction of working capital	Advanced sewage technology	Advanced sewage sludge technology
Scenarios 1	15	with	with	without	without
Scenarios 2	15	with	with	with	without
Scenarios 3	15	with	with	with	with
Scenarios 4	25	with	with	with	with

References

Beijing Municipal Bureau of Statistics (2011) Beijing Input-Output Extension Table 2010. BMBS, Beijing

Beijing Municipal Development and Reform Commission (2011) The Twelfth Five-Year Plan for the National Economic and Social Development of Beijing [EB/OL]: http://www.bjpc.gov.cn/fzgh_1/guihua/12_5/Picture_12_F_Y_P/, 21 Jan 2011

Beijing Municipal Bureau of Statistics (2012) Beijing Statistical Yearbook 2011. BMBS, Beijing

Chae SR, Shin HS (2007) Effect of condensate of food waste (CFW) on nutrient removal and behaviours of intercellular materials in a vertical submerged membrane bioreactor (VSMBR). Bioresour Technol 98(2):373–379

Higano Y, Sawada T (1997) The dynamic policy to improve the water quality of Lake Kasumigaura. Stud Reg Sci 26(1):75–86

Higano Y, Yoneta A (1999) Economic policies to relieve contamination of Lake Kasumigaura. Stud Reg Sci 29(3):205–218

Hirose F, Higano Y (2000) A simulation analysis to reduce pollutants from the catchment area of Lake Kasumigaura. Stud Reg Sci 30(1):47–63

Jing DB et al (2009) COD, TN and TP removal of Typha wetland vegetation of different structures. Pol J Environ Stud 18(2):183–190

Kyou HL, Jong H, Tae JP (1998) Simultaneous organic and nutrient removal from municipal wastewater by BSACNR process. Korean J Chem Eng 15(1):9–14

Li H, Jin YY, Li YY (2011) Carbon emission and low-carbon strategies of sewage sludge treatment. J Civ Archit Environ Eng 33(2):117–121

Mizunoya T, Sakurai K, Kobayashi S, Piao SH, Higano Y (2007) A simulation analysis of synthetic environment policy: effective utilization of biomass resources and reduction of environmental burdens in Kasumigaura basin. Stud Reg Sci 36(2):355–374

Wang YH, Inamori R, Kong HN, Xu KQ, Inamori Y, Kondoc T, Zhang JX (2008) Influence of plant species and wastewater strength on constructed wetland methane emissions and associated microbial populations. Ecol Eng 32:22–29

Yan JJ, Xao RG, Sha JH (2010) Comprehensive evaluation of integrated pollutant-minimization policies in rural areas of China. China Popul Resour Environ 20(3):124–127

Zhang GF, Xu F, Wang TY et al (2016) Comprehensive evaluation of the regional environmental and economic impacts of adopting advanced technologies for the treatment of sewage sludge in Beijing. In: Thill J-C (ed) Spatial analysis and location modeling in urban and regional systems in advances in spatial science. Springer, New York. Accepted 29 Oct 2013

Chapter 4
Comprehensive Simulation Results Analysis

Abstract In this simulation, we select Beijing to make empirical research. The economic and environmental impacts of sewage sludge treatment in each scenario are shown in this chapter. The optimal scenario is selected by comparing the simulation results. Especially, we make a regional analysis to evaluate the capacity of sustainable development in every subregion.

Keywords Simulation results · Optimal scenario · Subregion analysis

4.1 Economic Effects

4.1.1 Objective Function

Feasible solution was obtained in every scenario (Fig. 4.1). The value of the objective function (sum of GRP for eleven years) for Scenarios 1, 2, 3, and 4 are 21,118, 24,122, 24,417, and 24,151 billion CNY, respectively. Sum of GRP for eleven years for Scenarios 3 and 4 are more than of that for Scenarios 1 and 2. The result of the more production for Scenarios 3 and 4 is resulted from the introduction of both advanced sewage and sewage sludge treatment technology, which dispose of more water pollutants and allow industries not only more productions but also more water pollutants emissions.

4.1.2 Change in GRP from 2010 to 2020

Figure 4.2 shows the GRP growth trend from 2010 to 2020 for every scenario. In this simulation, the average rate of the increase in GRP of Beijing for ten years is 4.70 %, 7.94 %, 8.72 %, and 8.03 % in Scenarios 1, 2, 3, and 4, respectively. The simulation result in Scenario 1 demonstrates that constraints resulting from high

© The Author(s) 2016

G. Zhang, *Environmental and Social-economic Impacts of Sewage Sludge Treatment*, SpringerBriefs in Economics, DOI 10.1007/978-981-287-948-6_4

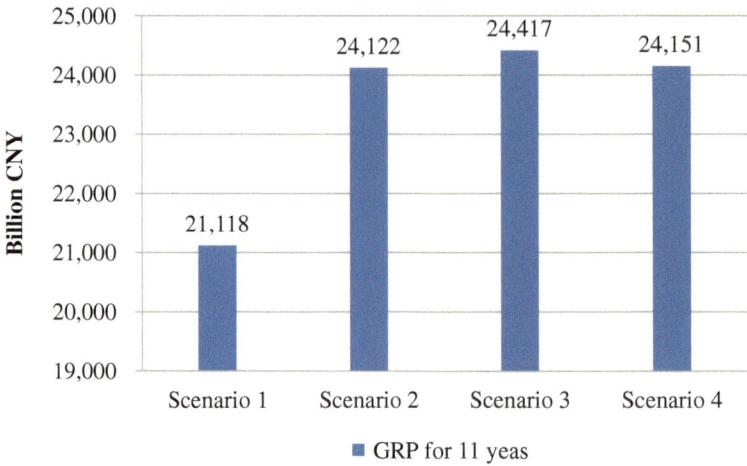

Fig. 4.1 Objective function (sum of GRP for 11 years)

levels of water pollutants will limit economic development. Furthermore, the GRP will decrease after 2018 in Scenario 1 in which without any new technology introduced. However, if we introduce advanced sewage treatment technologies, as in Scenario 2, the GRP growth rate is demonstrated by the simulation to reach a higher level. However, the increase in the GRP is greatest when both advanced sewage and sewage sludge treatment technologies are adopted, as in Scenario 3. In this scenario, the GRP in 2020 is more than twice that of 2011. It should be noted here that the rate for Scenario 4 is the second largest in amount of the four Scenarios, which also achieve the government economic development target.

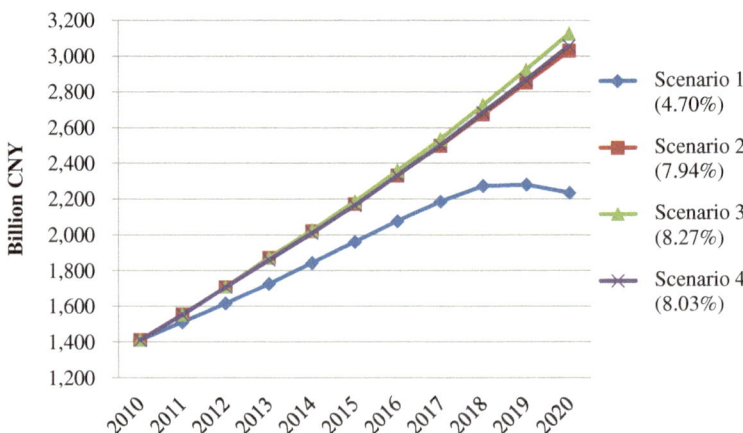

Fig. 4.2 GRP growth trend from 2010 to 2020 for every scenario

4.1.3 The Economic Impact of Technology Investment

In this section, we defied a cost-benefit ratio (α) to access the economic impact of technology investment.

$$\alpha = \frac{\Delta GRP}{\Delta S}, \qquad (4.1)$$

where ΔS is the total investment of new technology for eleven years; ΔGRP is the changes in GRP bought by new technology investment.

This ration is an important indicator for policymaking. And it reflects the economic feasibility of implementing the proposed policies. If α is more than one, it means that the investment is effective to improve economic development.

Figure 4.3 shows the calculation result based on the simulation. In Scenarios 2, 3, and 4, it indicates per unit technology investment increases will bring 300, 330, 303 CNY GRP rise, respectively; this demonstrates the benefits brought by new technology introduction.

Comparing Scenarios 2, 3, and 4, we find that the ratio in Scenarios 3 and 4 is higher than of that in Scenario 2. The reason is that new sewage sludge treatment technology introduced in these two scenarios. Although in Scenario 4 the ratio is lower than in Scenario 3 for the more strictly constraint of water pollutants, it is higher than of that in Scenario 2 where only introduces sewage treatment technology.

4.2 Environmental Impact

4.2.1 Load of Water Pollutants

Load of T–P for eleven years for Scenarios 1, 2, 3, and 4 are 54.46, 54.42, 53.19, and 49.84 thousand tons, respectively; load of T–N for eleven years for Scenarios 1, 2, 3, and 4 are 595, 592, 486, and 449 thousand tons, respectively; load of COD for

Fig. 4.3 Economic impact of technology investment

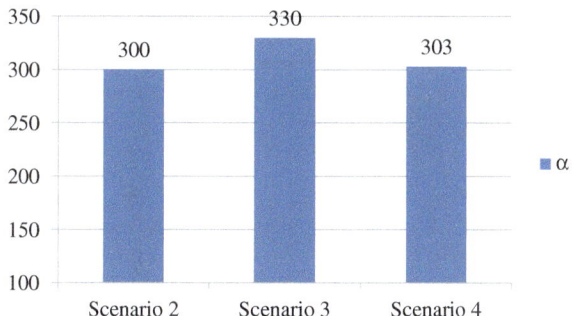

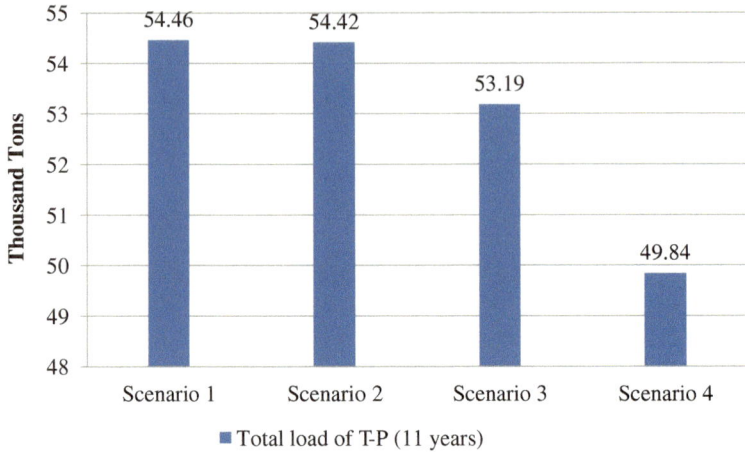

Fig. 4.4 Load of T–P for 11 years for every scenario

eleven years for Scenarios 1, 2, 3, and 4 are 1,896, 1,208, 2,046, and 1,933 thousand tons, respectively (Figs. 4.4, 4.5, and 4.6). These results show that load of T–P and T–N for Scenarios 1 and 2, where advanced sewage sludge treatment technologies are not adopted, are lower than of that for Scenarios 3 and 4, where advanced sewage sludge treatment technologies are adopted. By contrast, load of COD for Scenarios 1 and 2 is higher than of that for Scenarios 3 and 4. The reason is that COD has become the boundary of water pollutants constriction when we adopt sewage sludge treatment technology.

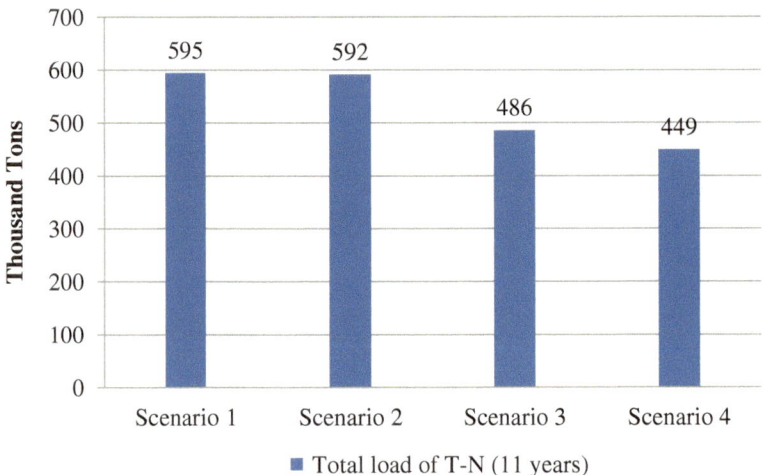

Fig. 4.5 Load of T–N for 11 years for every scenario

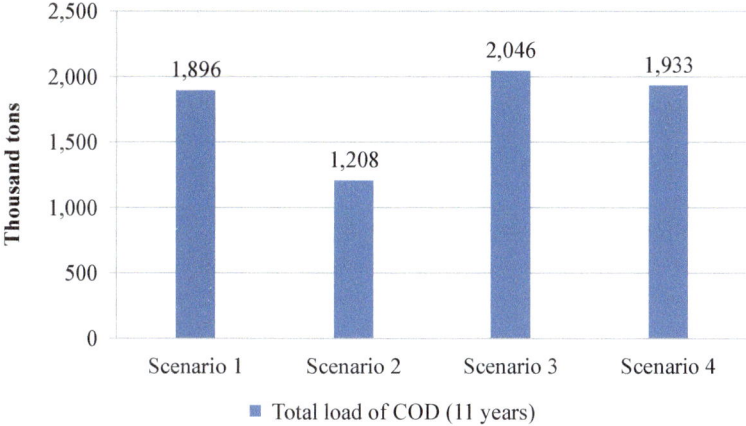

Fig. 4.6 Load of COD for 11 years for every scenario

4.2.2   GHG Emission Impact

(1) GHG Emission Intensity

GHG emission is determined by energy consumption. The simulation result shows that GHG emission intensity for Scenarios 1, 2, 3, and 4 are 35, 48, 47, and 47 thousand tons/billion CNY, respectively (Fig. 4.7). Those are also lower than the target value in 2020 of Beijing government's plan, where GHG emission intensity should be reduced by 36 % in 2020 compared with 2010 (the target value in 2020 is 50 thousand tons/billion CNY). This simulation result demonstrates that the constraint on water pollutants is stricter than the constraint on GHG emission intensity.

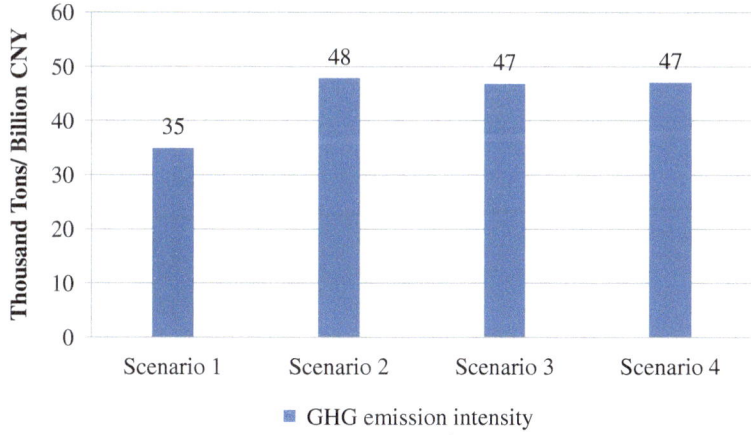

Fig. 4.7 GHG emission intensity in 2020 for every scenario

(2) GHG Emission of Sewage and Sewage Sludge

Figure 4.8 shows GHG emission of sewage and sewage sludge for eleven years. In this simulation, the total amount of GHG emission of sewage and sewage sludge of Beijing for eleven years is 35, 47, 28, and 28 million tons in Scenarios 1, 2, 3, and 4, respectively. A comparison of the results from Scenarios 1 and 2 demonstrate that with construction of new sewage plants, sewage treatment consumes more energy in Scenario 2 than that of in Scenario 1 which causes more GHG emission. In the same time, sewage sludge as the byproduct of sewage treatment is increasing. However, the capacity of sewage sludge treatment does not change. Therefore, the amount of untreated sewage sludge is also increased, which increased GHG emissions. However, in Scenarios 3 and 4 when we introduce both new sewage and sewage sludge treatment technology, the total amount of GHG emission by sewage and sewage sludge is less than that of in Scenario 1 in which without introduction of new technology and Scenario 2 in which only introduces sewage treatment technology. Because more sewage sludge is treated by new sewage sludge plants, the amount of GHG emission by untreated sewage sludge in Scenarios 3 and 4 is much less than of that in Scenarios 1 and 2.

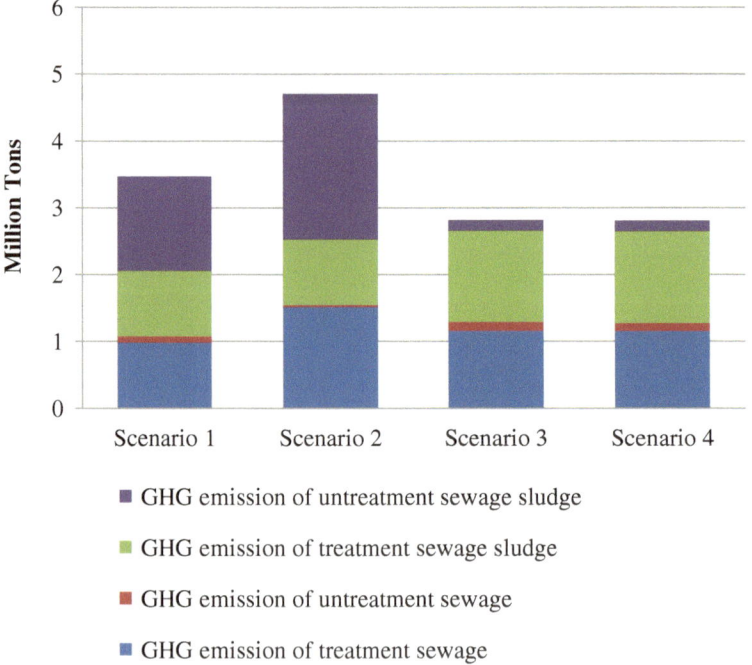

Fig. 4.8 GHG emission of sewage and sewage sludge for 11 years

Table 4.1 Cost of GHG reduction

Scenario	Scenario 2	Scenario 3	Scenario 4
ΔGHG (thousands ton)	320,371	327,824	298,967
ΔGRP (billion CNY)	3,004	3,298	3,033
β (thousands ton/billion CNY)	107	99	98

This simulation result also indicated that GHG emission reduction potential by proper treatment of sewage and sewage sludge is 1.9 million tons for eleven years. (3) Cost of GHG reduction
In this section, we defied the cost of GHG reduction (β) to access the economic efficiency of the integrated policy.

$$\beta = \frac{\Delta GRP}{\Delta GHG}, \tag{4.2}$$

where ΔGHG is the change of total GHG emission in Scenarios 2, 3, and 4 comparing with in Scenario 1; ΔGHP is the changes in GRP in Scenarios 2, 3 and 4 comparing with in Scenario 1.

Table 4.1 shown the calculate result of based on the simulation. The cost of GHG reduction in Scenarios 2, 3, and 4 is 107, 99, and 98 thousands ton/billion CNY, respectively. This result demonstrates that the integrated policy with introducing advanced sewage treatment technology in Scenarios 3 and 4 is more economic efficiency for reduction of GHG than of that without introducing the advanced sewage treatment technology in Scenario 2.

4.3 Energy Impact

4.3.1 Energy Consumption Intensity

According to the plan of Beijing government (Beijing Municipal Development and Reform Commission 2011), a 34 % reduction of energy consumption intensity (with energy consumption intensity is less than 33 thousands TCE/billion CNY in 2020) in 2020 was set up, compared with that of 2010. Based upon this simulation result, energy consumption intensity for Scenarios 1, 2, 3, and 4 are 25, 32, 32, and 32 thousands TCE/billion CNY, respectively. Those are lower than energy consumption intensity target in 2020 (Fig. 4.9). The reduction target of energy consumption intensity in 2020 is achievable for every scenario. This simulation result demonstrates that the constraint on water pollutants is stricter than the constraint on energy consumption intensity.

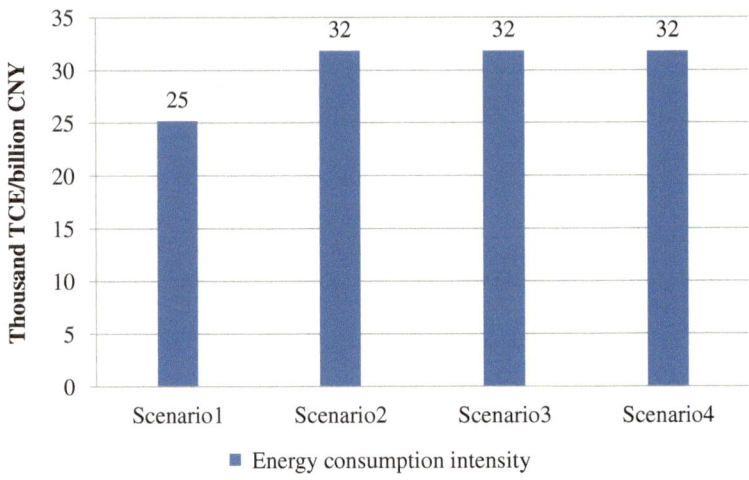

Fig. 4.9 Energy consumption intensity in 2020 for every scenario

4.3.2 Energy Consumption and Generation of Sewage Sludge Treatment

The energy consumption and generation of sewage sludge treatment for eleven years is shown in Fig. 4.10. In Scenarios 1 and 2, there is no sewage sludge plants construction. Energy consumption of sewage sludge treatment is stable at 510 thousands TCE for eleven years, and there is no energy production in the process of sewage sludge treatment because no advanced technology introduced. In Scenarios 3 and 4, with the new construction of sewage sludge plants, the amount of energy consumption by sewage sludge treatment is almost three times as in Scenarios 1 and 2. However, we should note that energy of 742 thousands TCE can be obtained as byproducts for eleven years when we introduce advanced sewage treatment technology, which account for about 48 % of energy consumption of sewage sludge treatment.

4.4 Optimal Case Selection

It is important to propose suitable policies and technologies for regional economic and environmental sustainable development, which need to conclude all the major factors that will impact on water environment, economic development, energy consumption, and GHG emission. In this simulation, we use four indicators based on the government plan to help us select the optimal case (Table 4.2). These indicators are water pollutants reduction rate, average growth rate of GRP, GHG

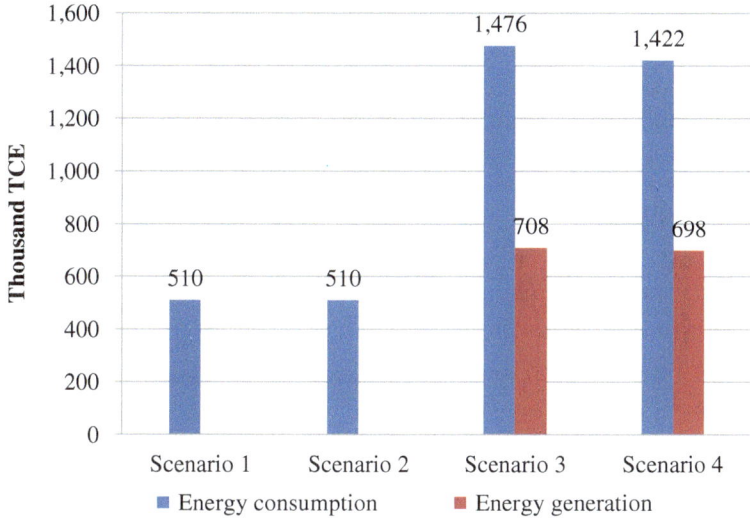

Fig. 4.10 Energy consumption and generation of sewage sludge treatment for every scenario for 11 years

emission intensity target in 2020, and energy consumption intensity target in 2020. Specifically, water pollutants reduction rate is equal or more than 15 % in 2020 compared in 2010; the average GRP growth rate for ten years is equal or more than 8 %; GHG emission intensity reduction rate is equal or more than 36 % in 2020 compared with in 2010; energy consumption intensity reduction rate is equal or more than 34 % in 2020 compared with in 2010.

Based on the analysis above, we can choose Scenario 4 as the optimal case for the achieving of the all government target and the largest reduction rate of water pollutants.

Table 4.2 Comparison of evaluation indicators for every scenario

Scenarios	Water pollutant reduction rate (%)	GRP growth rate is achievable	GHG emission intensity target is achievable	Energy intensity target is achievable
Scenario 1	15	No	Yes	Yes
Scenario 2	15	No	Yes	Yes
Scenario 3	15	Yes	Yes	Yes
Scenario 4	25	Yes	Yes	Yes

4.5   Regional Analysis

In this section, we will analyze the economic and environmental impacts in each subregion under the conditions of optimal scenario. Based on this simulation result, we confirmed that uniform policy may not get the identical impact (Siebert 1985), and the capacity of sustainable development is imbalance for these subregions.

4.5.1   Economic Development

The integrated policy is efficiency for economic growth of subregion. Figure 4.11 shows the GRP of subregions in 2010 and 2020. The trend of economic development shows that the GRP of each subregion is almost double the value in 2020 compared whit in 2010. Moreover, in terms of the GRP size of each region, the GRP of the Central City subregion accounts for about 75 % of total GRP in Beijing each year (Fig. 4.12a, b). This denotes that the Central City subregion plays an important role in the economic development of Beijing. In addition, we find that the proportion of GRP of Central City increased and Yanqing County decreased; and other subregions keep stable, by comparing in 2010 and 2020. The reason of this phenomenon is that the tertiary industry in Central City is the highest and in Yanqing County is the lowest in 2010 (Fig. 4.13).

4.5.2   Water Pollutant Intensity

Based upon the analysis shown above, the COD constraint becomes the target of interest for scenario 4. Therefore, in this section, we select COD intensity as a proxy for water environmental impact; this metric is an important indicator of environmentally sustainable regional economic development. Figure 4.14 shows the average COD intensity of every subregion from 2010 to 2020. Based on this simulation result, in Central City, Huairou, Mentougou, Fangshan, Penggu, Daxing, and Shunyi, the value of average COD intensity is lower than of that in Beijing in 2010 (143 tons/billion CNY). The reason for this phenomenon is that Beijing government supported subsidy for Central City, Huairou, Mentougou, Fangshan, and Penggu, to reduce water pollutants by construction of new sewage plant, and the economic development of Daxing and Shunyi is better than other subregions in 2010 (Table 4.3 and Fig. 4.11). This simulation also demonstrates that the capacity of water environmental sustainable development for every subregion is different.

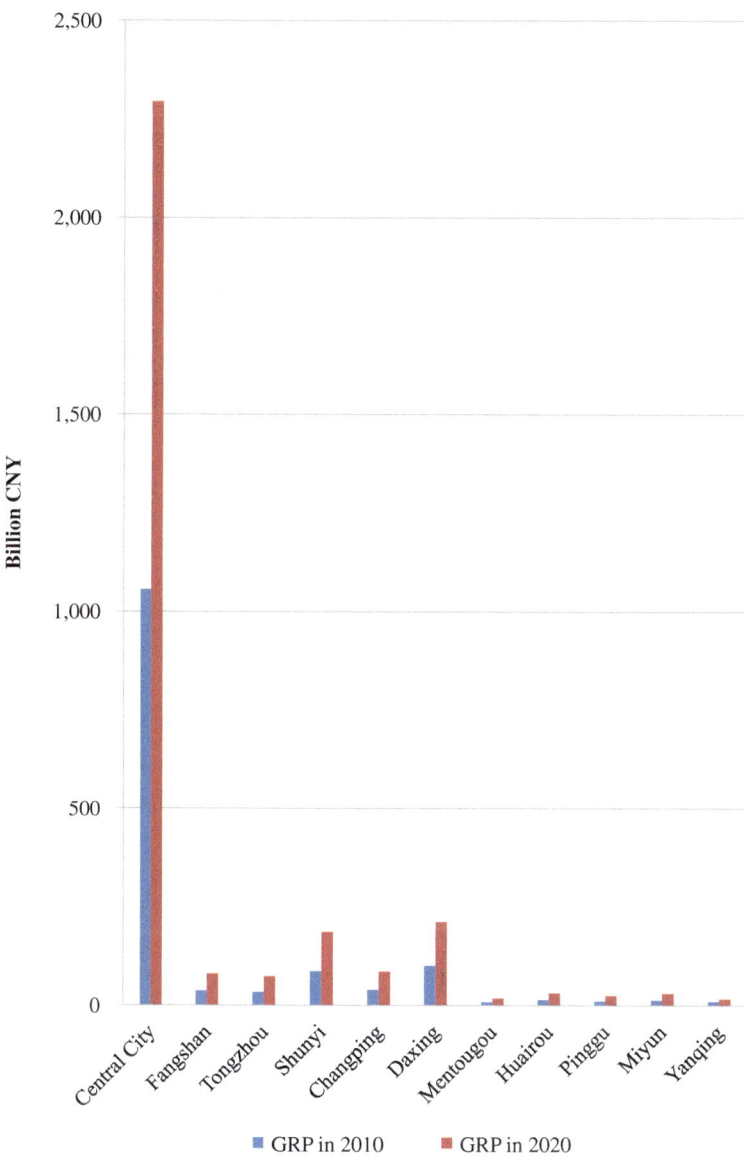

Fig. 4.11 GRP of every subregion in 2010 and 2020

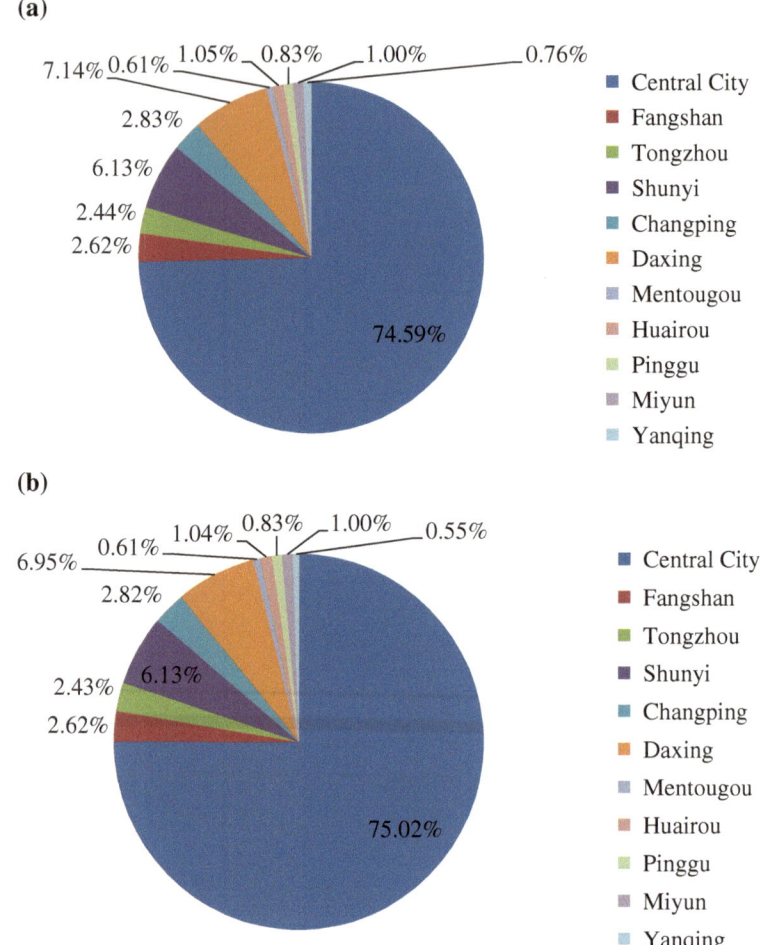

Fig. 4.12 **a** Proportion of the GRP for every subregion in 2010, **b** Proportion of the GRP for every subregion in 2020

4.5.3 Energy Consumption Intensity for Every Subregion in 2020

There is imbalance development for these subregions. Figure 4.15 shows energy consumption intensity for every subregion in 2020. Based on this simulation result, the values of energy consumption intensity for Shunyi, Central City, and Daxing are lower than the Beijing city' target (33 thousand TCE/billion CNY). However, based on the analysis above, when seen as a whole region, Beijing city can achieves this target. This simulation result also demonstrates that Central City, Shunyi, and Daxing are important subregions for energy consumption controlling.

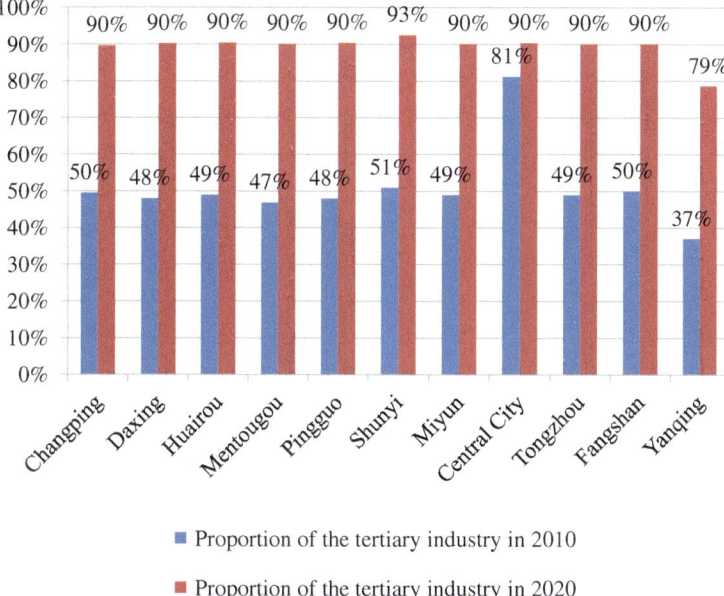

■ Proportion of the tertiary industry in 2010

■ Proportion of the tertiary industry in 2020

Fig. 4.13 The proportion of tertiary industry for every subregion in 2010 and 2020

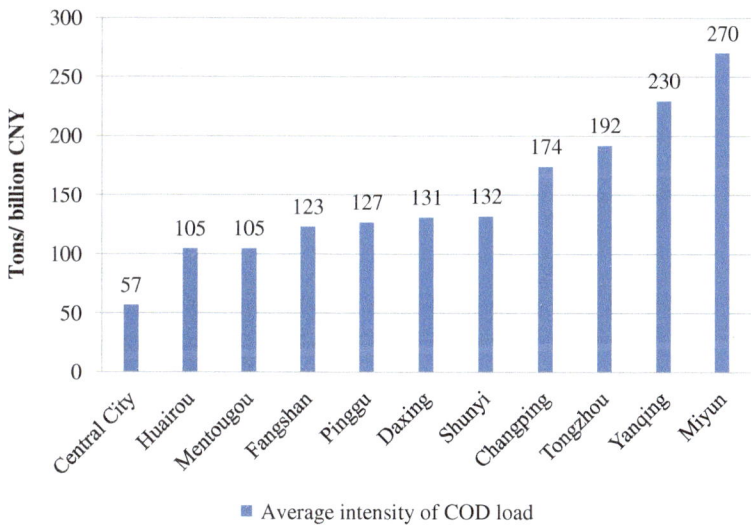

■ Average intensity of COD load

Fig. 4.14 Average COD intensity form 2010–2020 for every subregion

Table 4.3 Total subsidy for 11 years for every subregion in millions

Subregion	Subsidy for construction of new sewage plant	Subsidy for construction of new sewage sludge plant	Total	Percentage (%)
Central city	3,060	4,059	7,119	71.19
Fangshan	31	453	484	4.84
Tongzhou	0	453	453	4.53
Shunyi	0	404	404	4.04
Changping	0	429	429	4.29
Daxing	0	176	176	1.76
Mentougou	38	168	206	2.06
Huairou	39	206	245	2.45
Pinggu	23	218	240	2.40
Miyun	0	146	146	1.46
Yanqing	0	97	97	0.97
Total	3,191	6,809	10,000	100

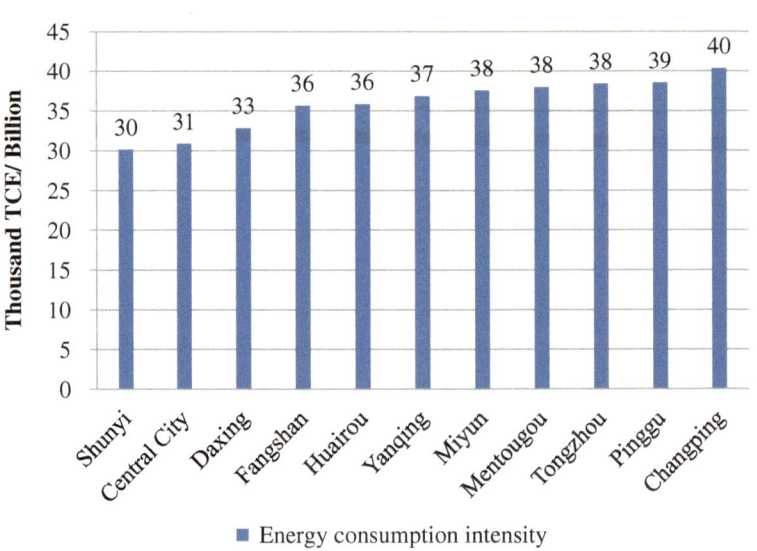

Fig. 4.15 Energy consumption intensity in 2020 in TCE/billion CNY

4.5.4 GHG Emission Intensity for Every Subregion in 2020

The GHG emission intensity is also uneven for these subregions. In this research, GHG emission is determined by energy consumption. Figure 4.16 shows GHG emission intensity for every subregion in 2020. Based on the simulation result,

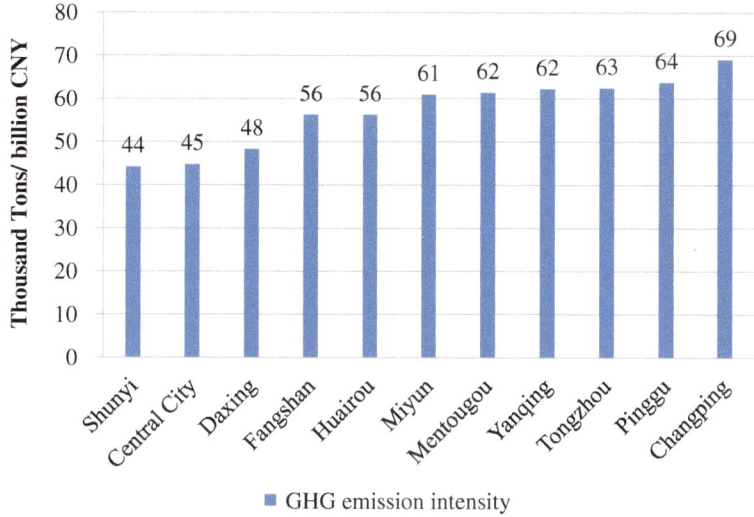

Fig. 4.16 GHG emission intensity in 2020 in TCE/billion CNY

Shunyi, Central City, and Daxing will achieve Beijing city' target (50 thousand tons/billion CNY) in 2020, because in this three subregion, the economic development situation is better than other subregion in 2010 (Fig. 4.11).

According to the analysis of economic and the environmental impact of subregions (Figs. 4.11, 4.12, 4.13, 4.14, and 4.15), we found that the unbalanced regional sustainable development exists. Uniform policies do not always lead to identical subregional impacts (Siebert 1985).

4.5.5 Optimal Subsidy for Every Subregion

Table 4.3 presents the optimal subsidy scheme for new sewage and sewage sludge plant construction. The subsidy consists of investments in new plant construction and maintenance costs. This simulation result assumes that the subsidy for the Central City subregion is in the greatest among all the subregions; this subsidy accounts for 71.20 % of the total subsidies for sewage and sewage sludge plant construction in Beijing. Because population and GRP of the Central City subregions accounts for 60 and 75 % of the total population and GRP of Beijing, respectively, in 2010. Although the rate of sewage treatment reached 95 % in 2010, a large amount of new sewage is generated as the population and GRP continue to grow.

In addition, because of the subsidy limitation for sewage treatment, only Central City, Fangshan, Mentougou Huairou, and Pinggu get finance support for their high-water quality requirements. But every subregion gets subsidy for sewage sludge treatment for achieving the government's requirements that all the sewage sludge should be treated in 2015.

4.5.6 New Sewage and Sewage Sludge Plants Construction Plan

The number of new sewage plant that is proposed to be constructed over eleven years (2010–2020) is shown in Table 4.4. MBR and DMBR, UMBR, EMBR represent the advanced technologies that will be used in the new sewage plants. A comparison for these sewage treatment technologies, DMBR and UMBR, is not used for its high investment and operational cost. And five subregions get the finance support for the high level of water environmental requirements under the constriction of subsidy. Because of the different of capacity between technology MBR and EMBR, Central City constructs nineteen plants with a focus on technology MBR and EMBR; Mentougou constructs three plants with technology EMBR; Huairou and Pinggu construct one plant with technology MBR, respectively.

The number of new sewage sludge plants that are proposed to be constructed over eleven years (2010–2020) is shown in Table 4.5. A–D–F-I, A–D–F-II, F–C-I, and F–C-II represent the advanced technologies that will be used in the new sewage sludge plants.

In this simulation, nine sewage sludge plants with the A–D–F-I technology, fifteen sewage sludge plants with the A–D–F-II technology and one sewage sludge plant with the F–C-II technology will need to be constructed in Central City subregion.

A comparison of this four kinds of new technologies, sewage sludge plants with A–D–F-I and A–D–F-II are mainly constructed for their low investment and

Table 4.4 New sewage plants construction in each region

Subregion	MBR	DMBR	UMBR	EMBR	Total
Center city	8	0	0	11	19
Fangshan	1	0	0	0	1
Tongzhou	0	0	0	0	0
Shunyi	0	0	0	0	0
Changping	0	0	0	0	0
Daxing	0	0	0	0	0
Mentougou	0	0	0	3	3
Huairou	1	0	0	0	1
Pinggu	1	0	0	0	1
Miyun	0	0	0	0	0
Yanqing	0	0	0	0	0
Total	11	0	0	14	25

Table 4.5 New sewage sludge plants construction in each region

	A–D–F-I	A–D–F-II	F–C-I	F–C-II	Total
Center city	5	6	0	1	12
Fangshan	0	1	0	0	1
Tongzhou	0	1	0	0	1
Shunyi	1	0	0	0	1
Changping	1	1	0	0	2
Daxing	0	1	0	0	1
Mentougou	1	1	0	0	2
Huairou	0	1	0	0	1
pinggu	0	1	0	0	1
Miyun	0	1	0	0	1
Yanqing	1	1	0	0	2
Total	9	15	0	1	25

operation cost. Only one plant with technology F–C-II is constructed in Central City as it can produce power directly. Technology A–D–F-II is almost used in every subregion its higher efficiency of GHG reduction. And A–D–F-I is used in Central City, Shunyi, Changping, Mentougou, and Yanqing for its lowest operation cost.

4.5.7   Sewage and Sewage Sludge Treatment Trend for Every Subregions

Figure 4.17 show the sewage and sewage sludge treatment rate of Beijing city from 2010 to 2020. This simulation results demonstrate that with implementation of the integrated policy, the capacity of sewage and sewage sludge treatment is increased year by year. In 2020, the rate of sewage treatment will achieve 88 %. However, in the view point of subregion, the rate of sewage treatment not always increases (Table 4.6). Some of subregions, such as Central City, Mentougou, Fangshan, Hairou, and Pinggu, get the subsidy for sewage treatment in which the speed of sewage treatment capacity increased is higher than the sewage generated. And because of the amount of sewage generation decreasing and stable sewage treatment capacity, the rate of sewage treatment will increase in Tongzhou, Shuiyi, and Yanqing in 2020 compared with in 2010. However, in Changping, Daxing, and Miyun, the rate of sewage treatment will decrease for the stable sewage treatment capacity, while the sewage generation is increasing.

We also should note that in order to realize the government plan, we set all the sewage to be treated in Beijing city from year 2011. Therefore, with the subsidy of sewage sludge treatment and construction of new sewage sludge plants, the rate of sewage sludge treatment is 99 % for every subregion from 2011 to 2020.

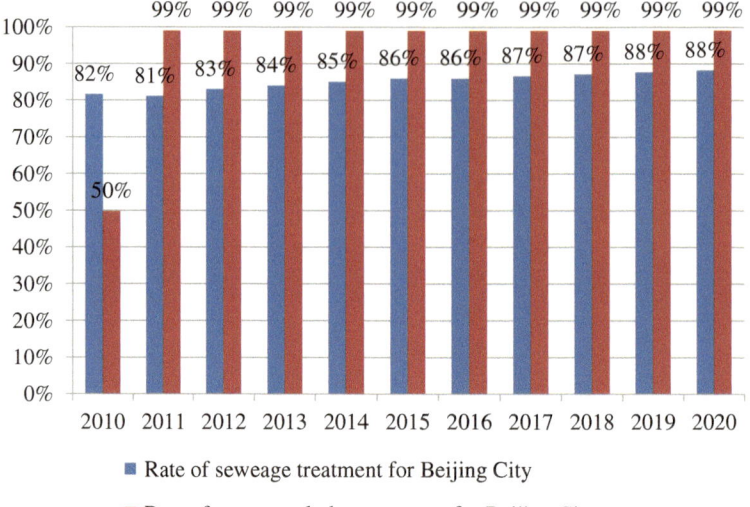

Fig. 4.17 Rate of sewage and sewage sludge treatment from 2010 to 2020

Table 4.6 Sewage generation and treatment capacity in 2010 and 2020 for subregion	Region	Capacity of sewage treatment (thousand million)		Sewage generation (thousand million)		Rate of sewage treatment (%)	
		2010	2020	2010	2020	2010	2020
	Central city	779	939	810	953	96	99
	Fangshan	36	51	52	57	69	90
	Tongzhou	51	51	84	61	61	84
	Shunyi	40	40	79	75	51	53
	Changping	90	90	120	124	75	73
	Daxing	68	68	95	108	72	63
	Mentougou	14	17	40	17	35	98
	Huairou	9	24	22	24	41	100
	Pinggu	19	34	32	35	59	98
	Miyun	16	16	46	48	35	33
	Yanqing	11	11	45	18	24	62
	Beiijing	1,133	1,341	1,425	1,520	80	88

4.5.8 Potential for Sustainable Development

The sustainable development ability is different for these subregions. In this study, we select four indicators to evaluate it. These indicators are intensity of COD load (load of COD per billion GRP), average growth rate of GRP, GHG emission intensity target in 2020, and energy consumption intensity target in 2020. Specifically, intensity of COD load is equal or more than 143 tons per billion GRP which is the value of COD intensity in Beijing city in 2010; the average GRP growth rate for ten years is equal or more than 8 %; GHG emission intensity is equal or less than 50 thousand tons per billion GRP; energy consumption intensity is equal or less than 33 thousands TCE per billion GRP. Moreover, each indicator gives equal weight.

Based on the simulation result, the sustainable development ability of subregion is shown in Table 4.7. Central City and Shunyi have the highest sustainable development ability for its high-economic growth rate and low-environmental pollution and energy consumption. Yanqing and Tongzhou have the lowest sustainable development ability for the low economic growth rate and high environmental pollution and energy consumption.

4.6 Discussion and Conclusions

In this study, we create an integrated (economic-environment) model and use a computer simulation to analyze the regional environmental and economic impacts of adopting advanced technologies for the treatment of sewage sludge. Moreover, we discuss the issues of economic development, environmental conservation and compensation, energy consumption intensity, and the ability of sustainable development of each subregion. The following conclusions are based upon the simulation results.

The integrated policy of promotion of water conservation, reduction of working capital, and the introduction of advanced sewage and sewage sludge treatment technologies is effective to reduce environmental pollutants and achieve economic development. In the optimal scenario (Scenario 4), the total GRP for 2010–2020 is 24,151 billion CNY and the average rate of economic growth from 2010 to 2020 is 8.03 %. Moreover, the total net load of T–P, T–N and COD is 49 thousand tons, 449 thousand tons, and 1,933 thousand tons, respectively, in Scenario 4. The reduction rate of T–P, T–N and COD is 46 %, 40 %, and 25 %, respectively, in 2020 compared with 2010 by keeping the target of 8 % GRP growth.

If the reduction rate of COD is 25 % in 2020 compared with in 2010, the reduction rate of energy consumption intensity and GHG emission intensity are 39 % and 36 %, respectively, both of which can achieve the government's plan. Moreover, GHG emission reduction potential by proper treatment of sewage and

Table 4.7 Comparison of evaluation indicators for every subregion

Subregion	Economic growth rate > 8 %	Intensity of COD load < 143tons/billion CNY	Energy consumption intensity < 32,625 TCE/billion CNY	GHG emission intensity < 49,953 tons/billion CNY	Ranking (high to low)
Central city	Yes	Yes	Yes	Yes	1
Shunyi	Yes	Yes	Yes	Yes	1
Daxing	No	Yes	Yes	Yes	2
Fangshan	Yes	Yes	No	No	3
Mentougou	Yes	Yes	No	No	3
Pinggu	Yes	Yes	No	No	3
Changping	Yes	No	No	No	4
Huairou	No	Yes	No	No	4
Miyun	Yes	No	No	No	4
Yanqing	No	No	No	No	5
Tongzhou	No	No	No	No	5

sewage sludge is 1.9 million tons for eleven years. And energy of 724 thousands TCE can be obtained as byproducts for eleven years.

The optimal budget expenditures for the policy are 3.19 billion CNY for new sewage plant construction and 6.81 billion CNY for new sewage sludge plant construction for eleven years. In specific, the optimal subsidy for sewage and sewage sludge plants construction of every subregion for eleven years is that: for Central City, Fangshan, Tongzhou, Shunyi, Changping, Daxing, Mentougou, Huairou, Pinggu, Miyun, and Yanqing is 7,120 million CNY, 484 million CNY, 453 million CNY, 404 million CNY, 429 million CNY, 176 million CNY, 206 million CNY, 245 million CNY, 240 million CNY, 146 million CNY, 97 million CNY, respectively.

The optimal sewage and sewage sludge plants construction plan is that: construct eleven new sewage plants featuring the MBR technology and fourteen plants featuring the EMBR technology; nine sewage sludge plants with the A–D–F-I technology, fifteen sewage sludge plants with the A–D–F-II technology and one sewage sludge plant with the F–C-II technology.

In this study, we also confirm that uniform policies do not always lead to identical subregional impacts (Siebert 1985). Unbalanced regional sustainable development exists. By comparing the simulation results, Central City (zone 1) plays a key point of sustainable development of Beijing city. The size of GRP will be about 75 % of total GRP of Beijng City in the next eleven years, while water pollutants intensity is lower than the other subregions.

References

Beijing Municipal Development and Reform Commission (2011) The Twelfth Five-Year Plan for the National Economic and Social Development of Beijing. [EB/OL]: http://www.bjpc.gov.cn/fzgh_1/guihua/12_5/Picture_12_F_Y_P/, 21 Jan 2011

Siebert H (1985) Spatial aspects of environmental economics. In: Kneese AV, Sweeney JL (eds) Handbook of natural resource and energy economics. Elsevier, Amsterdam, pp 125–164

Chapter 5
Conclusion

Abstract This chapter summarizes the main conclusions. The topics of summary of finding includes that analysis of current situation, feasibility of evaluation sewage sludge treatment with advanced technology, comprehensive evaluation of socio-economic and environmental policies with using advanced technology for sewage sludge treatment. In addition, potential use of this study for policy-making is briefly discussed. Finally, some suggestions for further work are provided.

Keywords Finding · Suggestions · Further work

5.1 Summary of Findings

5.1.1 Analysis of Current Situation

From the current situation analysis of Beijing city, we can see that improper treatment of sewage and sewage sludge is an important reason that causes water pollution.

About 2.8 billion tons sewage and 0.6 million tons sewage generated without any treatment which seriously pollutes the water environment. In 2010, water shortages and water pollution in urban river downstream serious situation still has not been fundamentally reversed. The river water quality in Beijing is bad compared with other types of water class; 43.2 % of the river water is considered to be "inferior V" class, which is the worst level in China. 6.3 % of lake water is considered to be "inferior V" class.

Recently, the government has realized the importance of environmental protection. Beijing government plans that all sewage sludge be treated by 2015 and load of chemical oxygen demand (COD) be reduced by 8.7 % in 2015 compared with 2010.

Therefore it is necessary to treat sewage and sewage sludge to improve water environment. Also, it is necessary to research the economic impact of sewage and

© The Author(s) 2016

G. Zhang, *Environmental and Social-economic Impacts of Sewage Sludge Treatment*, SpringerBriefs in Economics, DOI 10.1007/978-981-287-948-6_5

sewage sludge treatment by using advanced treatment technology. Furthermore, it is beneficial to determine the optimal development plan for Beijing to realize regional sustainable development.

5.1.2   Comprehensive Evaluation of Socio-Economic and Environmental Policies with Using Advanced Technology for Sewage Sludge Treatment

From this comprehensive simulation, we compared the economic and environmental impact of the integrated policies with an emphasis on sewage sludge treatment in different circumstances.

In the optimal scenario (Scenario 4), the total GRP for 2010–2020 is 24,151 billion CNY and the average rate of economic growth from 2010 to 2020 is 8.03 %. Moreover, the total net load of T-P, T-N and COD is 49 thousand tons, 449 thousand tons, and 1,933 thousand tons, respectively, in Scenario 4. The reduction rate of T-P, T-N and COD is 46, 40 and 25 %, respectively, in 2020 compared with 2010 by keeping the target of 8 % GRP growth.

In the optimal Scenario the reduction rate of energy consumption intensity and GHG emission intensity reach more than 39 and 36 %, respectively, both of which can achieve the government's plan. Moreover, GHG emission reduction potential by proper treatment of sewage and sewage sludge is 1.9 million tons for eleven years. And energy of 724 thousands TCE can be obtained as byproducts for eleven years.

Regional analysis provides a more specific plan for Beijing city development, including sewage and sewage sludge plants construction and subsidy distribution. The ability of sustainable development of sub-regions analyses show that Central City and Shunyi have the highest ability of sustainable development. Central City (zone 1) plays a key point of sustainable development of Beijing city. The size of GRP will be about 75 % of total GRP of Beijing City in the next eleven years while water pollutants intensity is lower than the other sub-regions.

5.2   Potential Use of This Study for Policy-Making

Simulation results indicated that the integrated policy of promotion of water conservation, reduction of working capital and the introduction of advanced sewage and sewage sludge treatment technologies is effective to reduce environmental pollutants and achieve economic development. This simulation also proves that advanced technology introduce is significant for achieving the dual targets of economic development and environment protection.

This simulation results also can help government to make policy for constructing sewage and sewage sludge plant, selecting technology, distributing financial budget and so on.

5.3   Suggestion for Further Work

This study focuses only on organic pollution measures (T-P, T-N and COD) commonly used in the literature to describe water pollution. As it is important for public health, other toxic waste, such as heavy metal, should be investigated further, though this is outside of the research scope.

In this study, analysis on Beijing economic system function is a closed sub-national region. We conducted the model with a short-run eleven years prediction. Our ultimate goal is to reconstruct the model as an open system using a long-run perspective, extending the study area to encompass north China, and eventually a national or global study area. In that case we would consider both the pollution flow and the economic relations among sub-regions.

Appendix A

See (Table A.1).

Table A.1 Sewage treatment plant of Beijing in 2010

District	The name of sewage treatment plants	Main processes	Running time	Design processing capacity (Million m3 / day)	Average processing capacity (Million m3 / day)
Chaoyang	Gaobeidian sewage treatment plant	A/A/O	1999	100	77.54
	Jiuxianqiao sewage treatment plant	Oxidation ditch	2000	20	20.3
	Xiaohongmen sewage treatment plant	A/A/0	2006	60	58.4
	Beiyuan sewage treatment plant	Oxidation ditch	2009	4	3.6
	Beixiaohe sewage treatment plant	A/A/0	1990	10	7.56
Fengtai	Wujiacun sewage treatment plant	A/A/0	2003	8	6.06
	Fangzhuang sewage treatment plant	A/A/0	1995	4	4
	Fengtaigangsewage treatment plant	Biochemistry	1996	1.2	1.1
Shijingshan	Liugou bridge sewage treatment plant	A/A/0	2004	10	6.3
Haidian	Yongfeng sewage treatment plant	Oxidation ditch	2008	2	1.73
	Wenquan sewage treatment plant	A/A/0	2008	2	1.14

(continued)

Table A.1 (continued)

District	The name of sewage treatment plants	Main processes	Running time	Design processing capacity (Million m3 / day)	Average processing capacity (Million m3 / day)
Haidian	Qinghe sewage treatment plant	A/A/0	2004	40	45
	Xiaojiahe sewage treatment plant	A/A/0	2003	2	2
Mentougou	Mencheng sewage treatment plant	SBR	2004	4	1.1
Fangshan	Changyang sewage treatment plant	CASS	2008	2	0.5
	Chengguan sewage treatment plant	Oxidation ditch	2009	2	1.73
	Liangxiang sewage treatment plant	A/0	2003	4	3.5
	Noukouyu sewage treatment plant	Oxidation ditch	1995	6	2.5
Tongzhou	Bishui sewage treatment plant	Biochemistry	2005	10	8.1
	Guoxian sewage treatment plant	Oxidation ditch	2009	1	0.2
	Zhangjiawan sewage treatment plant	Activated sludge		1	0.5
	Ciqu sewage treatment plant	Oxidation ditch	2005	1	0.5
Shunyi	Tianzhu sewage treatment plant	MHA	2006	2	1.2
	Tongshun sewage treatment plant	Biochemistry	2008	1	0.4
	Shunyi sewage treatment plant	Oxidation ditch	2007	8	7.45
Changping	Beiqijia river sewage treatment plant	CASS	2008	2.5	2.45
	Nankou town sewage treatment plant	Oxidation ditch	2009	2	1.1
	Tiantongyuan sewage treatment plant	SBR	2009	1.32	1.26
	Chuangping sewage treatment plant	Oxidation Ditch	2003	5.4	2.4
	Xiaotangshan sewage treatment plant	CASS	2006	1.05	0.6

(continued)

Table A.1 (continued)

District	The name of sewage treatment plants	Main processes	Running time	Design processing capacity (Million m3 / day)	Average processing capacity (Million m3 / day)
Daxing	Xingshui sewage treatment plant	Oxidation ditch	2000	1.1	0.25
	Tiantanghe sewage treatment plant	A/A/0	2009	5	3.8
	Panggezhuang sewage treatment plant	Oxidation ditch	2009	7.5	4.66
	Jingyuan sewage treatment plant	MBR	2002	5	3.8
Huairou	Huairou sewage treatment plant	Oxidation ditch	2007	7.5	4.66
Pinggu	Jiahe sewage treatment plant	A/A/0	2008	8	4.6
Miyun	Tanzhou sewage treatment plant	Biochemistry	2001	4.5	2.4
Yanqing	Xiadujingyan sewage treatment plant	SBR	2006	3	1.8

Source http://wenku.baidu.com/view/2ab2279c51e79b89680226cc.html

Appendix B

See (Table B.1).

Table B.1 Industry output in Beijing in 2010

	Industry	Final consumption	Capital formation	Net export	Net transfer	Total output
1	Agriculture	18.16	2.68	−6.79	−12.48	15.42
2	Forestry	2.32	0.34	0.87	−1.59	1.97
3	Animal husbandry	16.61	2.45	−6.21	−11.42	14.10
4	Fishery	1.55	0.23	−0.58	−1.06	1.31
5	Minerals mining	1.71	3.24	−222.78	188.25	109.39
6	Processing of petroleum, coking and processing of nuclear fuel	78.05	15.52	−1.00	−91.97	130.40
7	Chemical industry	13.44	3.66	−8.39	−37.72	115.72
8	Metallurgical industry	3.79	12.68	7.40	−141.76	107.56
9	Iron and steel industry	25.29	70.94	45.38	62.21	409.67
10	Manufacture of communication equipment, computers and other electronic equipment	12.40	17.23	22.61	−93.57	252.23
11	Other manufacturing	4.38	−0.67	1.17	−3.80	13.37
12	Production and distribution of electric power and heat	22.84	−2.66	18.04	−54.79	300.64
13	Construction	7.08	239.98	21.44	−18.51	316.98
14	Transportation, warehousing and postal service	9.62	3.47	−83.97	−7.75	253.38
15	Other services	573.55	240.63	47.18	494.99	2520.03

© The Author(s) 2016
G. Zhang, *Environmental and Social-economic Impacts of Sewage Sludge Treatment*, SpringerBriefs in Economics, DOI 10.1007/978-981-287-948-6

Appendix C

See (Table C.1).

© The Author(s) 2016

G. Zhang, *Environmental and Social-economic Impacts of Sewage Sludge Treatment*, SpringerBriefs in Economics, DOI 10.1007/978-981-287-948-6

Table C.1 Input–output coefficient of 15 industries of Beijing in 2010

	1	2	3	4	5	6	7	8	9	10	11	12	13	14	15
1	0.0809	0.0809	0.0809	0.0809	0.0001	0.0377	0.0090	0.0000	0.0000	0.0000	0.0004	0.0000	0.0011	0.0000	0.0019
2	0.0103	0.0103	0.0103	0.0103	0.0000	0.0048	0.0012	0.0000	0.0000	0.0000	0.0000	0.0000	0.0001	0.0000	0.0002
3	0.0740	0.0740	0.0740	0.0740	0.0001	0.0345	0.0083	0.0000	0.0000	0.0000	0.0003	0.0000	0.0010	0.0000	0.0018
4	0.0069	0.0069	0.0069	0.0069	0.0000	0.0032	0.0008	0.0000	0.0000	0.0000	0.0000	0.0000	0.0001	0.0000	0.0002
5	0.0395	0.0395	0.0395	0.0395	0.4197	0.0104	0.0401	0.2430	0.0055	0.0010	0.0696	0.1247	0.0461	0.0050	0.0011
6	0.0198	0.0198	0.0198	0.0198	0.0101	0.2666	0.0177	0.0063	0.0084	0.0049	0.0613	0.0052	0.0236	0.0065	0.0295
7	0.0463	0.0463	0.0463	0.0463	0.0093	0.0574	0.2804	0.0306	0.0305	0.0357	0.0300	0.0069	0.0219	0.0017	0.0268
8	0.0144	0.0144	0.0144	0.0144	0.0116	0.0337	0.0234	0.3023	0.0882	0.0318	0.2759	0.0078	0.3266	0.0103	0.0110
9	0.0159	0.0159	0.0159	0.0159	0.0194	0.0096	0.0121	0.0171	0.4078	0.0262	0.0051	0.0168	0.0738	0.0642	0.0282
10	0.0007	0.0007	0.0007	0.0007	0.0054	0.0019	0.0032	0.0023	0.0343	0.4964	0.0019	0.0031	0.0039	0.0009	0.0597
11	0.0003	0.0003	0.0003	0.0003	0.0011	0.0020	0.0032	0.0184	0.0017	0.0024	0.1185	0.0007	0.0067	0.0003	0.0017
12	0.0708	0.0708	0.0708	0.0708	0.1412	0.0309	0.1104	0.0334	0.0130	0.0096	0.0216	0.4805	0.0399	0.1275	0.0324
13	0.0000	0.0000	0.0000	0.0000	0.0013	0.0011	0.0021	0.0003	0.0005	0.0006	0.0006	0.0014	0.0219	0.0039	0.0229
14	0.0399	0.0399	0.0399	0.0399	0.0953	0.0526	0.0726	0.0724	0.0335	0.0289	0.0615	0.0589	0.0440	0.3944	0.0570
15	0.2013	0.2013	0.2013	0.2013	0.0908	0.2150	0.1500	0.1174	0.1708	0.2426	0.1274	0.0983	0.1924	0.1282	0.3306